U.S.NRC

United States Nuclear Regulatory Commission

Protecting People and the Environment

NUREG-1555, Supplement 1
Revision 1

I0493577

Standard Review Plans for Environmental Reviews for Nuclear Power Plants

Supplement 1: Operating License Renewal

Final Report

Office of Nuclear Reactor Regulation

AVAILABILITY OF REFERENCE MATERIALS
IN NRC PUBLICATIONS

NRC Reference Material

As of November 1999, you may electronically access NUREG-series publications and other NRC records at NRC's Public Electronic Reading Room at http://www.nrc.gov/reading-rm.html. Publicly released records include, to name a few, NUREG-series publications; *Federal Register* notices; applicant, licensee, and vendor documents and correspondence; NRC correspondence and internal memoranda; bulletins and information notices; inspection and investigative reports; licensee event reports; and Commission papers and their attachments.

NRC publications in the NUREG series, NRC regulations, and Title 10, "Energy," in the *Code of Federal Regulations* may also be purchased from one of these two sources.
1. The Superintendent of Documents
    U.S. Government Printing Office
    Mail Stop SSOP
    Washington, DC 20402–0001
    Internet: bookstore.gpo.gov
    Telephone: 202-512-1800
    Fax: 202-512-2250
2. The National Technical Information Service
    Springfield, VA 22161–0002
    www.ntis.gov
    1–800–553–6847 or, locally, 703–605–6000

A single copy of each NRC draft report for comment is available free, to the extent of supply, upon written request as follows:
Address: U.S. Nuclear Regulatory Commission
         Office of Administration
         Publications Branch
         Washington, DC 20555-0001
E-mail: DISTRIBUTION.RESOURCE@NRC.GOV
Facsimile: 301–415–2289

Some publications in the NUREG series that are posted at NRC's Web site address http://www.nrc.gov/reading-rm/doc-collections/nuregs are updated periodically and may differ from the last printed version. Although references to material found on a Web site bear the date the material was accessed, the material available on the date cited may subsequently be removed from the site.

Non-NRC Reference Material

Documents available from public and special technical libraries include all open literature items, such as books, journal articles, transactions, *Federal Register* notices, Federal and State legislation, and congressional reports. Such documents as theses, dissertations, foreign reports and translations, and non-NRC conference proceedings may be purchased from their sponsoring organization.

Copies of industry codes and standards used in a substantive manner in the NRC regulatory process are maintained at—
    The NRC Technical Library
    Two White Flint North
    11545 Rockville Pike
    Rockville, MD 20852–2738

These standards are available in the library for reference use by the public. Codes and standards are usually copyrighted and may be purchased from the originating organization or, if they are American National Standards, from—
    American National Standards Institute
    11 West 42nd Street
    New York, NY 10036–8002
    www.ansi.org
    212–642–4900

NUREG-1555, Supplement 1
Revision 1

United States Nuclear Regulatory Commission

Protecting People and the Environment

Standard Review Plans for Environmental Reviews for Nuclear Power Plants

Supplement 1: Operating License Renewal

Final Report

Manuscript Completed: May 2013
Date Published: June 2013

Office of Nuclear Reactor Regulation

ABSTRACT

This document provides guidance to U.S. Nuclear Regulatory Commission staff in implementing the provisions in Title 10 of the *Code of Federal Regulations* Part 51 (10 CFR Part 51), "Environmental Protection Regulations for Domestic Licensing and Related Regulatory Functions" when conducting an environmental review for the renewal of a nuclear power plant operating license(s). This standard review plan guides the staff in preparing a plant-specific supplemental environmental impact statement to NUREG-1437, Revision 1, *Generic Environmental Impact Statement for License Renewal of Nuclear Plants*. This document supplements NUREG-1555, *Standard Review Plans for Environmental Reviews for Nuclear Power Plants*, which provides guidance for the environmental reviews of construction permits, initial operating licenses, early site permits, and combined licenses for new nuclear power plants.

PAPERWORK REDUCTION ACT STATEMENT

CONTENTS

ABBREVIATIONS AND ACRONYMS

APE   area of potential effects

CEQ   Council on Environmental Quality
CFR   *Code of Federal Regulations*
CWA   Clean Water Act

DBA   design-basis accident

EFH   essential fish habitat
EIS   environmental impact statement
EMF   electromagnetic field
EPA   U.S. Environmental Protection Agency
EPM   environmental project manager
ER   environmental report
ESA   Endangered Species Act
ESRP   environmental standard review plan (NUREG-1555)
ESRP/S1   environmental standard review plan, Supplement 1 (NUREG-1555, Supplement 1)

FES   final environmental statement
FR   *Federal Register*
FWPCA   Federal Water Pollution Control Act

GEIS   *Generic Environmental Impact Statement for License Renewal of Nuclear Power Plant* (NUREG-1437)

MSA   Magnuson-Stevens Act

NEPA   National Environmental Policy Act of 1969
NESC   National Electrical Safety Code
NHPA   National Historic Preservation Act
NMFS   National Marine Fisheries Service
NPDES   National Pollutant Discharge Elimination System
NRC   U.S. Nuclear Regulatory Commission
NRHP   *National Register of Historic Places*
NRR   Office of Nuclear Reactor Regulation

PILOT   payments in lieu of tax
PUD   Public Utility District

ROW(s)   right(s) of way
ROI   region of influence, interest, or impact

SAMA	severe accident mitigation alternatives
SAMDA	severe accident mitigation design alternatives
SEIS	supplemental environmental impact statement
SHPO	State Historic Preservation Officer
THPO	Tribal Historic Preservation Officer
USFWS	U.S. Fish and Wildlife Service

INTRODUCTION

This environmental standard review plan (ESRP) consists of a series of instructions developed for U.S. Nuclear Regulatory Commission (NRC) staff use in conducting environmental reviews for the renewal of nuclear power plant operating licenses and preparing plant-specific supplemental environmental impact statements (SEISs) to NUREG-1437, Revision 1, the *Generic Environmental Impact Statement for License Renewal of Nuclear Plants* (GEIS). This ESRP amends NUREG-1555, Supplement 1, "Standard Review Plans for Environmental Reviews for Nuclear Power Plants: Environmental Standard Review Plan for Operating License Renewal," issued October 1999. Use of this ESRP helps ensure the completeness and consistency of the environmental review and analyses conducted by the NRC staff in its SEISs. This ESRP is a companion to Regulatory Guide 4.2, Supplement 1, Revision 1, *Preparation of Environmental Reports for Nuclear Power Plant License Renewal Applications.*

Questions regarding the content of any plan in this document may be directed to the NRC at the following address:

> Environmental Review and Guidance Update Branch
> Division of License Renewal
> Office of Nuclear Reactor Regulation
> U.S. Nuclear Regulatory Commission
> Washington, DC 20555-0001

Additional copies of these plans may be obtained as indicated on the inside front cover of this document.

NRC's Implementation of the National Environmental Policy Act

This ESRP demonstrates how NRC staff meets the provisions in Title 10 of the *Code of Federal Regulations* (10 CFR) Part 51, "Environmental Protection Regulations for Domestic Licensing and Related Regulatory Functions" to conduct environmental reviews for the renewal of nuclear power plant operating licenses and prepare plant-specific SEISs to the GEIS. The NRC regulations at 10 CFR Part 51 implement Section 102(2) of the National Environmental Policy Act (NEPA). The NRC published the license renewal provisions of 10 CFR 51 in the *Federal Register* on December 18, 1996 (61 FR 66537), which became effective on January 17, 1997. The NRC's intention in developing the rule was to improve the efficiency of the environmental review process for the renewal of nuclear power plant operating licenses. These provisions also support the analyses conducted for and reported in the GEIS.

Environmental Review Process

After receiving an application for license renewal, including the applicant's environmental report (ER), the NRC staff performs an acceptance review of the ER to determine whether the information provided is sufficiently complete to begin the environmental/NEPA review process. After reviewing the information and assessments presented in the applicant's ER, the NRC staff begins to prepare the plant-specific SEIS. These ESRPs guide the NRC staff's environmental review and preparation of the plant-specific SEIS. In the SEIS, the NRC staff analyzes the environmental impacts of renewing the operating license of a nuclear power plant (the proposed action) and the alternatives to renewing the license. The completed plant-specific SEIS presents the staff's recommendation whether to renew the operating license of the nuclear power plant. The NRC's record of decision (ROD) considers this recommendation, along with the findings from the safety review (10 CFR Part 54).

The NRC's NEPA review process consists of the following actions required by 10 CFR Part 51:

- Publish a notice of intent to prepare an SEIS in the *Federal Register* (see 10 CFR 51.27, "Notice of Intent," and 10 CFR 51.95(c), "Post Construction Environmental Impact Statements: Operating License Renewal Stage") and send copies of the notice to appropriate Federal, State, and local agencies; Indian Tribes; appropriate State, regional, and metropolitan clearinghouses; and any interested persons upon request (see 10 CFR 51.116, "Notice of Intent"). The notice describes the proposed action and explains the scoping process, states the locations of copies of the ER available for public inspection, and invites public participation in the scoping process.

- Conduct scoping (see 10 CFR 51.28, "Scoping—Participants," 10 CFR 51.29, "Scoping—Environmental Impact Statement and Supplement to Environmental Impact Statement," and 40 CFR 1506.6(b)(3), "Public Involvement"). The scoping process includes identifying and inviting the appropriate agencies, Indian Tribes, interest groups, and persons to participate in the process. Concerning license renewal, scoping focuses on allowing other parties to raise environmental issues that they believe are significant and yet are not addressed or not adequately addressed in the ER. Parties may raise issues at the public scoping meeting, which the NRC staff routinely holds in the vicinity of the plant, and in written comments. The scoping process also routinely includes a staff visit to the plant site and communication with local, regional, and State officials and representatives of interested or knowledgeable organizations. As a result of scoping, the staff may request additional information from the applicant.

- Prepare a draft SEIS (see 10 CFR 51.70, "Draft Environmental Impact Statement—General," 10 CFR 51.71, "Draft Environmental Impact Statement—Contents," and 10 CFR 51.95(c)). In developing the draft SEIS, the NRC staff will evaluate (verify and validate) information provided by the applicant and will seek out and collect information from independent sources.

- Distribute the draft SEIS for comment (see 10 CFR 51.73, "Request for Comments on Draft Environmental Impact Statement"). The NRC will publish a notice of the availability of the SEIS in the *Federal Register* and will distribute copies of the draft SEIS to the U.S Environmental Protection Agency (EPA); appropriate Federal, State, and local agencies; Indian Tribes; appropriate State, regional, and metropolitan clearinghouses; organizations and individuals who have expressed interest in the review; and any other parties requesting a copy (see 10 CFR 51.74, "Distribution of Draft Environmental Impact Statement and Supplement to Draft Environmental Impact Statement; News Releases").

- Prepare a final SEIS (see 10 CFR 51.95(c)). In developing the final SEIS, the NRC staff will consider comments received on the draft, prepare responses, and modify the SEIS as warranted. This includes determining whether the comments identify new and significant information that was neither considered in the GEIS nor addressed in the applicant's ER. After considering the environmental impacts associated with license renewal and replacement power alternatives, the staff will determine whether or not the adverse environmental impacts of license renewal are so great that preserving the option of license renewal for energy planning decisionmakers would be unreasonable. The NRC will publish a notice of the availability of the final SEIS in the *Federal Register* and will distribute copies to the U.S Environmental Protection Agency; appropriate Federal, State, and local agencies; Indian Tribes; appropriate State, regional, and metropolitan clearinghouses; organizations and individuals who have expressed interest in the review and participated in the environmental review; and any other parties requesting a copy (see 10 CFR 51.93, "Distribution of Final Environmental Impact Statement and Supplement to Final Environmental Impact Statement; News Releases," and 10 CFR 51.118, "Final Environmental Impact Statement—Notice of Availability").

- Hold a hearing on the license renewal application if the Commission or the designated licensing board determines that it is in the public interest or if a request for hearing and petition to intervene are granted. In accordance with 10 CFR 2.105(a)(10), "Notice of Proposed Action," the agency will issue a notice of opportunity for hearing as soon as practicable after the application has been docketed. Any person whose interest may be affected by the action may request a hearing. (See also 10 CFR 51.104, "NRC Proceeding Using Public Hearings; Consideration of Environmental Impact Statement.")

- Publish a ROD (see 10 CFR 51.103, "Record of Decision—General"). Among other things, the ROD will discuss the alternatives considered in the SEIS, the measures taken to minimize environmental harm, and any license conditions adopted in connection with mitigation measures. In making a final decision on license renewal, the NRC will determine whether or not the adverse environmental impacts of license renewal are so great that preserving the option of license renewal for energy planning decisionmakers would be unreasonable. The NRC publishes the Commission's final decision on the application in the *Federal Register*.

For license renewal, the environmental project manager (EPM) is responsible for the environmental review and preparation of the plant-specific SEIS. The EPM interacts with the applicant's technical and supervisory personnel as well as with NRC management. In addition, the EPM coordinates the efforts of technical staff and contractor personnel during the acceptance review of the applicant's ER and the environmental review conducted for the plant-specific SEIS. With assistance from the technical staff, the EPM develops the overall recommendations for action to be taken by the Director of the Office of Nuclear Reactor Regulation (NRR).

EPM responsibilities include managing the acceptance review of the ER and the environmental review conducted for the plant-specific SEIS. The purpose for the acceptance review is to determine whether the applicant's ER provides sufficient information and analysis to support the plant-specific environmental review. If the ER is acceptable, it is docketed, and the environmental review may begin.

The environmental review is conducted by the environmental technical staff in NRR's Division of License Renewal, NRR's Division of Risk Assessment (DRA), and by the EPM. The responsibilities of the environmental technical staff in carrying out the environmental review, including criteria for ER acceptability, are provided in this ESRP. During the course of the technical staff's environmental review, it may be necessary to request additional information from the applicant. Requests for additional information are transmitted to the applicant by the EPM. RAIs also serve as a public record of the staff's concerns about the applicant's ER during the environmental review.

Environmental standard review plans for each resource area contained in this ESRP provide procedures for the environmental review leading to the preparation of the draft SEIS. The EPM is responsible for reviewing the draft SEIS and ensuring that the technical staff's conclusions meet NRC NEPA requirements and reflect NRC policy. In accordance with NRC standards, it is expected that each SEIS prepared by NRC staff will:

- stand on its own as an analytical document which fully informs decisionmakers and the public of the environmental effects of the proposed action and those of reasonable alternatives

- emphasize the issues that are significant and reduce emphasis on other issues and background material

- be written in plain language

The draft SEIS is submitted for review and comment to the division director, the Office of the General Counsel, and the division branch chiefs. Final approval is obtained from the EPM's division director before publication of the draft SEIS for public comment.

The Generic Environmental Impact Statement for License Renewal of Nuclear Plants (GEIS; NUREG-1437)

The 1996 GEIS[1] was prepared to address the environmental impacts associated with the renewal of nuclear power plant operating licenses. The GEIS identifies environmental impact issues that are common to all nuclear power plants, and impact issues that require plant-specific analyses. The NRC staff would prepare plant-specific SEISs to address the impact issues that could not be generically dispositioned in the GEIS.

The GEIS is intended to improve the efficiency of the license renewal process by (1) providing an evaluation of the types of environmental impacts that may occur from renewing commercial nuclear power plant operating licenses, (2) identifying and assessing impacts that are expected to be generic (the same or similar) at all nuclear plants (or plants with specified plant or site characteristics), and (3) defining the number and scope of environmental impact issues that need to be addressed in plant-specific SEISs.

The NRC committed to review and update the original findings in Table B-1, "Summary of Findings on NEPA Issues for License Renewal of Nuclear Power Plants," located in Appendix B to Subpart A of 10 CFR Part 51. The Commission stated that it intends to review the assessment of impacts in Table B-1 and the GEIS and update it on a 10-year cycle, if necessary. Since publication of the 1996 GEIS, over 40 nuclear plant sites (approximately 70 reactor units) have been the subject of plant-specific environmental reviews. The GEIS revision is intended to incorporate lessons learned and knowledge

---

1 Any reference in this document to the 1996 GEIS includes the two-volume set issued in 1996 and Addendum 1 to the GEIS issued in 1999.

gained from these plant-specific environmental reviews, as well as other new information and research published since the 1996 GEIS.

The NRC staff reviewed and reevaluated the environmental impact issues and findings in the 1996 GEIS. Experience and knowledge gained from license renewal reviews conducted since the 1996 GEIS provided new sources of information for the evaluations presented in the revision. In addition, the NRC staff considered new research, findings, and other information in evaluating the significance of impacts associated with license renewal. The purpose of the evaluation was to determine if the findings presented in the 1996 GEIS remained valid. In doing so, the NRC considered the need to modify, add to, or delete any of the 92 issues in the 1996 GEIS.

The revised GEIS evaluates 78 environmental impact issues; analyses determined that 60 of these issues are adequately addressed for all nuclear power plants. These issues are Category 1 issues, and do not require additional analysis in a plant-specific environmental review unless new and significant information is found. Of the remaining 18 environmental impact issues, 17 are Category 2 issues, which require plant-specific analyses. One issue (chronic effects of electromagnetic fields) is not categorized due to the lack of a scientific consensus on its impacts, and the NRC staff does not perform a plant-specific analysis of this issue in SEISs. Once a consensus has been reached by appropriate Federal health agencies on the potential health effects from electromagnetic fields, the NRC will revise its guidance and evaluation of this issue.

In the 1996 GEIS, a standard of significance was established for assessing environmental issues. Significance indicates the importance of likely environmental impacts and is determined by considering two variables: context and intensity. Context is the geographic, biophysical, and social context in which the effects will occur. In the case of license renewal, the context is the environment surrounding the facility. Intensity refers to the severity of the impact, in whatever context it occurs. The NRC developed a three-level standard of significance based upon the Council on Environmental Quality (CEQ) guidelines (40 CFR 1508.27):

- SMALL – environmental effects are not detectable or are so minor that they will neither destabilize nor noticeably alter any important attribute of the resource. For the purposes of assessing radiological impacts, the Commission has concluded that those impacts that do not exceed permissible levels in the Commission's regulations are considered small.

- MODERATE – environmental effects are sufficient to noticeably alter important attributes of the resource but not to destabilize them.

- LARGE – environmental effects are clearly noticeable and are sufficient to destabilize.

In addition to a determination of significance of environmental impacts associated with an issue, a determination was made whether the analysis in the GEIS could be applied to all nuclear plants (as well as to all plants with certain plant or site characteristics). Issues were assigned a Category 1 or Category 2 designation as follows:

Category 1 issues are those that meet all of the following criteria:

- The environmental impacts associated with the issue have been determined to apply either to all plants or, for some issues, to plants having a specific type of cooling system or other specified plant or site characteristics;

- A single significance level (i.e., small, moderate, or large) has been assigned to the impacts (except for collective off-site radiological impacts from the fuel cycle and from high-level waste and spent fuel);

- Mitigation of adverse impacts associated with the issue has been considered in the analysis, and it has been determined that additional plant-specific mitigation measures are not likely to be sufficiently beneficial to warrant implementation.

For issues that meet the three Category 1 criteria, no additional plant-specific analysis is required in future SEISs unless new and significant information is identified.

Category 2 issues are those that do not meet one or more of the criteria of Category 1, and, therefore, require additional plant-specific review.

Scope of the Environmental Standard Review Plans

The ESRPs in Supplement 1, Revision 1, guides the review of the environmental impact issues associated with license renewal. The ESRPs address all of the environmental impact issues discussed in the revised GEIS as well as any new environmental impact issues identified through the public scoping process. They also provide the framework for conducting impact analyses and preparing sections for the plant-specific SEIS. A review procedure is provided for each Category 2 issue. The ESRPs also provide for the systematic integration of new and significant information on Category 1 issues.

Use of the ESRPs in the environmental review process for license renewal would ensure

- identification of environmental impact issues, data and other information, and analysis

- consideration of specific environmental issues of concern to Federal, State, regional, and local agencies, and Indian Tribes, as appropriate

- standardization of review procedures for the analysis of environmental impact issues

- focused environmental review of potentially significant environmental impacts

Organization of the Environmental Standard Review Plans

The ESRPs are grouped into seven chapters. These chapters are as follows:
1. Purpose and Need for the Proposed Action
2. Alternatives Including the Proposed Action
3. Affected Environment
4. Environmental Consequences and Mitigating Actions

5. Environmental Impacts of Postulated Accidents

6. Alternatives to License Renewal

7. Summary and Conclusions

Chapters 1 through 3 are descriptive in nature. They guide the review of the purpose and need for the proposed action, the identification of reasonable alternatives to the proposed action, and the description of the nuclear plant site and the affected environment. ESRP Chapters 4 and 5 address the analysis of environmental impacts. They guide the review of the potential environmental impacts associated with continued plant operations and refurbishment associated with license renewal. ESRP Chapters 6 and 7 address the evaluation of the alternatives to license renewal. They guide the comparison of the proposed action with reasonable alternatives and the summarization of the conclusions regarding the environmental impacts of license renewal.

The environmental review plans in Chapters 4 and 5 identify Category 1 and 2 issues and new and significant information. Review plans serve to guide in the:

*   evaluation of the applicant's process for identifying and evaluating new information

*   evaluation of information submitted by members of the public during the scoping process, and information identified during the environmental review to determine whether new information is significant

*   identification of the information required to complete a plant-specific review of the issue

*   preparation of statements for the SEIS that describe the issue and present the conclusion

The format of the ESRPs in this document consists of the following six sections:

I.   Areas of Review

II. Acceptance Criteria

III. Review Procedures

IV. Evaluation Findings

V.   Implementation

VI. Bibliography

Areas of Review describes the purpose and scope of the environmental review. **Acceptance Criteria** provides guidance on determining the acceptability of the environmental impact analysis in the SEIS. **Review Procedures** describe the methods the staff uses in conducting the environmental review. The level of detail in the methods of environmental review varies from review plan to review plan. **Evaluation Findings** provides guidance on how to summarize the conclusions of the environmental review. **Implementation** describes how the review plan is used. Finally, the **Bibliography** section contains the bibliographic reference information supporting the material cited in the review plan.

Each ESRP provides a list of data and information needs under **Areas of Review**. The following sources of information should be considered:

- applicant's environmental report (ER)

- previous NRC Final Environmental Statements and other Environmental Impact Statements (e.g., SEISs)

- applicant's Safety Analysis Report or Updated Final Safety Analysis Reports

- NRC Safety Evaluation Reports

- Generic Environmental Impact Statement for License Renewal of Nuclear Plants, NUREG-1437, Volumes 1, 2, and 3, Revision 1.

- other Federal and State agencies

- the Internet and other online information

New and Significant Information

In the SEIS, the NRC staff is required to address any new and significant information on the environmental impacts of license renewal involving Category 1 and Category 2 issues. This section describes the identification of new information, evaluation of the significance of new information, and the treatment of new and significant information. When no new and significant information is found, a statement should be included in the SEIS that briefly describes the search for and evaluation of new information and states that no new information was identified or the new information was determined to be not significant.

The NRC staff must identify any new information on the environmental impacts of license renewal. The process for identifying new and significant information should consider:

- **The applicant's ER.** Applicants are required by 10 CFR 51.53(c)(3)(iv) to disclose new and significant information of environmental impacts of license renewal of which they are aware. In reviewing the applicant's ER, NRC staff must consider the applicant's process for discovering and evaluating the significance of any new information. Is the process adequate to ensure a reasonable likelihood that the applicant would be aware of new information, if it existed?

- **Records of public meetings and correspondence related to the application.** Compare information presented by the public with information considered in the GEIS. Is the information new in the sense that it post dates the analysis conducted for the GEIS?

- **Environmental quality standards and regulations.** Have the applicable environmental quality standards and regulations changed since the analysis conducted for the GEIS? If so, do the changes affect the NRC evaluation of applications for license renewal?

- **Technical literature.** Does recent technical literature contain information that would alter conclusions in the GEIS for Category 1 issues? Does the information indicate that there may be environmental impacts that were not considered in the GEIS?

Any new information should be used to develop precisely defined environmental impact issues. After the impact issues have been defined, the significance level of each issue should be determined using the significance level definitions in the GEIS. Appropriate mitigation measures should be identified and considered for each issue for which there is an adverse environmental impact. The consideration of mitigation measures should be in proportion to the potential adverse impact.

If the significance level is moderate or large, the reviewer should prepare an impact assessment for inclusion in the appropriate section of the plant-specific SEIS. The assessment should include a concise description of the new environmental impact information (including source) and how this information applies to the nuclear plant. The statement should also identify the significance level of the potential adverse impacts, and should list any mitigation measures that would be considered appropriate. A summary statement and a list of references cited in the impact assessment should also be provided.

General Instructions

The following instructions, applicable to most of the ESRPs, are provided here to avoid repetition in each plan:

- **Project Overview**. As an initial step in each individual environmental review, the reviewer is expected to develop an understanding of the proposed action. The purpose of this instruction is to ensure that reviewers put their individual reviews in perspective with the proposed action and concentrate their efforts on significant issues. This general project review is to be conducted as the first step (acceptance review phase) of the overall environmental review process and is to be completed before developing requests for additional information.

- **Internal Review Coordination**. The EPM is the central point of contact for all reviewers. Although each ESRP represents a discrete segment of NRC's environmental review, no review can be completed without coordination with related reviews. For example, the technical analysis ESRPs (in Chapters 4 and 5) rely on the descriptive chapters (1 through 3) for background information. All reviewers are instructed to maintain close communication with other reviewers throughout the review procedure. With few exceptions, the reviews are conducted in parallel; thus, other environmental reviews may not be available to reviewers before their own environmental review is completed.

- **External Review Coordination**. The EPM initiates contact with outside agencies and must be informed of all subsequent contacts made by reviewers. Each reviewer is expected to be aware of any related technical analyses and environmental assessments. Particular attention should be given to analyses and environmental assessments prepared under provisions of memoranda of understanding between the NRC and other Federal, State, regional, and local agencies, and Indian Tribes. Working through the EPM, the reviewer is responsible for resolving any differences of opinion between NRC staff analyses and analyses conducted by other agencies. The reviewer must ensure that all viewpoints are presented or that the specific provisions of the memoranda of understanding are followed.

- **Consultation with Other Agencies**. The environmental reviews may require consultation with other Federal, State, regional, and local agencies, and Indian Tribes. Federal agencies include, but are not limited to, the U.S. Fish and Wildlife Service and the National Marine Fisheries Service concerning threatened and endangered species, the State Historic Preservation Officer and Indian Tribes concerning historic and cultural resources listed or eligible for listing on the *National Register of Historic Places*; the Environmental Protection Agency (or designated State agencies) responsible for compliance with the Clean Water Act; State agencies responsible for consistency determinations under the Coastal Zone Management Act and State Implementation Plans under the Clean Air Act. These consultations should be started as soon as possible during the environmental review process and should be coordinated with the EPM.

- **Consultation with the Applicant**. All consultations or discussions with the applicant are made through the EPM.

- **Site Visit**. Most reviewers benefit from a visit to the nuclear plant site. This visit provides the reviewer with firsthand knowledge of site and the location and position of facilities. It also allows the reviewer an opportunity to study the environment around the nuclear plant site.

- **Depth of Review**. The reviewer must conduct an environmental impact analysis in sufficient depth to permit verification and validation of the analysis and conclusions.

- **Consideration of Mitigation**. Mitigation measures should be considered in proportion to the level of impact when adverse impacts are identified. Statements should also describe the potential effectiveness of mitigation measures.

- **Best Management Practices**. The reviewer must evaluate the applicant's commitments to use practices that minimize, reduce, or avoid adverse impacts. These practices, often referred to as best management practices, are activities that can mitigate potential adverse environmental impacts.

- **Quality Assurance**. Reviewers should identify and evaluate the quality assurance measures taken by the applicant in the collection and analysis of data. Quality assurance measures are also evaluated when computer models have been used to predict environmental impacts.

- **Findings**. Findings should reflect "consensus" agreement among the reviewers. This requires input from the reviewer, the EPM, and any other NRC reviewers affected by the findings.

- **Documentation**. Each reviewer should maintain documentation, logs, and other records of communication and consultation with outside agencies and organizations.

- **Definitions**. Use of the following terminology applies only to the environmental review process. Terms such as plant and station, used in an EIS, continue to reflect the choice of terms used to identify the nuclear plant (e.g., Calvert Cliffs Nuclear Power Plant, Oconee Nuclear Station).

    - Station: Consists of all facilities (reactor containment, turbine, and control buildings, intakes, discharges, etc.) located on the power plant site. Generally, the station includes everything located on the applicant's property that supports the existing reactor(s). In some cases, intake and discharge structures may be located offsite, but are considered part of the station. Transmission lines and their associated facilities are generally not considered part of the station. Other facilities not associated with the production of electricity (e.g., a visitor center or a fish hatchery), however, are considered part of the station.

- Plant: The nuclear reactor, reactor power conversion systems, intake and discharge structures, and all other facilities involved with the production of electricity. A plant can be more than one reactor and power conversion system. Transmission lines and other off-station facilities are not part of the plant.

- Unit: One reactor power conversion system. Generally, the term "unit" is used only when the applicant is proposing to relicense more than one unit.

- Facility: Any individual identifiable part of the station or plant. Examples: The visitor center is a facility. A substation is a facility. An intake system could be a facility (if discussed separately from the remainder of the plant).

- Mitigation: Impact mitigation is the process of modifying an activity to prevent, eliminate, and/or reduce the adverse environmental impact.

Bibliography

10 CFR Part 2. *Code of Federal Regulations*, Title 10, *Energy,* Part 2, "Rules of Practice for Domestic Licensing Proceedings and Issuance of Orders."

10 CFR Part 51. *Code of Federal Regulations*, Title 10, *Energy,* Part 51, "Environmental Protection Regulations for Domestic Licensing and Related Regulatory Functions."

10 CFR Part 54. *Code of Federal Regulations*, Title 10, *Energy,* Part 54, "Requirements for Renewal of Operating Licenses for Nuclear Power Plants."

40 CFR Part 1500. *Code of Federal Regulations*, Title 40, *Protection of the Environment*, Part 1500, "Purpose, Policy, and Mandate."

40 CFR Part 1501. *Code of Federal Regulations*, Title 40, *Protection of the Environment*, Part 1501, "NEPA and Agency Planning."

40 CFR Part 1502. *Code of Federal Regulations*, Title 40, *Protection of the Environment*, Part 1502, "Environmental Impact Statement."

40 CFR Part 1503. *Code of Federal Regulations*, Title 40, *Protection of the Environment*, Part 1503, "Commenting."

40 CFR Part 1504. *Code of Federal Regulations*, Title 40, *Protection of the Environment*, Part 1504, "Predecision Referrals to Council of Proposed Federal Actions Determined to be Environmentally Unsatisfactory."

40 CFR Part 1505. *Code of Federal Regulations*, Title 40, *Protection of the Environment*, Part 1505, "NEPA and Agency Decision Making."

40 CFR Part 1506. *Code of Federal Regulations*, Title 40, *Protection of the Environment*, Part 1506, "Other Requirements of NEPA."

40 CFR Part 1507. *Code of Federal Regulations*, Title 40, *Protection of the Environment*, Part 1507, "Agency Compliance."

40 CFR Part 1508. *Code of Federal Regulations*, Title 40, *Protection of the Environment*, Part 1508, "Terminology and Index."

61 FR 66537. U.S. Nuclear Regulatory Commission. Environmental Review for Renewal of Nuclear Power Plant Operating Licenses. Final Rule. December 18, 1996.

Clean Air Act of 1970, as amended. 42 USC 7401 et seq.

Clean Water Act (CWA). 33 USC 1251 et seq.

Coastal Zone Management Act of 1972 (CZMA), as amended. 16 USC 1451 et seq.

National Environmental Policy Act of 1969, as amended (NEPA). 42 USC 4321 et seq.

U.S. Nuclear Regulatory Commission (NRC). 1996. *Generic Environmental Impact Statement for License Renewal of Nuclear Plants*. NUREG-1437, Vols. 1 and 2. Office of Nuclear Reactor Regulation, Washington, D.C.

U.S. Nuclear Regulatory Commission (NRC). 1999. "Section 6.3—Transportation, Table 9.1, Summary of Findings on NEPA Issues for License Renewal of Nuclear Power Plants, Final Report." In *Generic Environmental Impact Statement for License Renewal of Nuclear Plants Main Report*. NUREG-1437, Vol. 1, Addendum 1. Washington, D.C.

U.S. Nuclear Regulatory Commission (NRC). 2013a. *Preparation of Environmental Reports for Nuclear Power Plant License Renewal Applications*. Regulatory Guide 4.2, Supplement 1, Revision 1.

U.S. Nuclear Regulatory Commission (NRC). 2013b. *Generic Environmental Impact Statement for License Renewal of Nuclear Plants*. NUREG-1437, Vols. 1, 2, and 3, Revision 1. Office of Nuclear Reactor Regulation, Washington, D.C.

U.S. NUCLEAR REGULATORY COMMISSION

ENVIRONMENTAL STANDARD REVIEW PLAN

OFFICE OF NUCLEAR REACTOR REGULATION

1.0 PURPOSE AND NEED FOR THE PROPOSED ACTION

I. AREAS OF REVIEW

This environmental standard review plan (ESRP) provides guidance for the preparation of the purpose and need for the proposed action. The discussion of purpose and need is found in Section 1.3 of the *Generic Environmental Impact Statement* (GEIS) *for License Renewal of Nuclear Plants* (NUREG-1437, Volumes 1, 2, and 3, Revision 1).

II. ACCEPTANCE CRITERIA

The reviewer should ensure that the introduction is consistent with the following regulations:

- 10 CFR 51.70(b). The draft environmental impact statement will be concise, clear, and analytic, and written in plain language with appropriate graphics....The format provided in Section 1(a) of Appendix A of this subpart should be used. The Nuclear Regulatory Commission (NRC) staff will independently evaluate and be responsible for the reliability of all information used in the draft environmental impact statement.

- 10 CFR 51.95(c), concerning the renewal of an operating license or combined license for a nuclear power plant. Under Parts 52 or 54 of this chapter, the Commission shall prepare an environmental impact statement, which is a supplement to the Commission's NUREG-1437, "Generic Environmental Impact Statement for License Renewal of Nuclear Plants."

- 10 CFR Part 51, Appendix A to Subpart A of Part 51, concerning format for presentation of material in environmental impact statements

- 10 CFR Part 51, Appendix A(4), concerning purpose of and need for action

Technical Rationale

Renewal of an operating license by the NRC is just one of the conditions required for continued safe operation of a nuclear power plant beyond the term of the initial license. Renewing the operating license would provide the licensee, State regulators, and utility officials with the option of extending plant operations beyond the term of the initial license should circumstances warrant it, whereas not renewing the operating license eliminates this option. Therefore, the Commission has defined the purpose and need for license renewal in terms of providing the licensee, State regulators, and utility officials with the option of extending the operating period of the nuclear plant. One or more introductory paragraphs should be prepared that present the Commission's definition of purpose and need.

III. REVIEW PROCEDURES

The material to be prepared is informational in nature; no specific analysis of the data is required.

IV. EVALUATION FINDINGS

The reviewer should prepare one or more introductory paragraphs for the supplemental environmental impact statement (SEIS) and should include the purpose and need for license renewal as it appears in Section 1.3 of the GEIS.

> The purpose and need for the proposed action (i.e., renewal of a commercial nuclear power plant operating license) is to provide an option to continue plant operations beyond the current operating license term to meet future system generating needs, as such needs may be determined by State, utility, system, and, where authorized, Federal (other than NRC) decisionmakers. Unless there are findings in the safety review required by the Atomic Energy Act or NEPA environmental review that would lead the NRC to reject a license renewal application, the NRC has no role in energy planning decisions of power plant owners, State regulators, system operators, and, in some cases, other Federal agencies, as to whether the plant should continue to operate.

V. IMPLEMENTATION

The method described in this ESRP would be used by the staff in evaluating conformance with the Commission's regulations, except in those cases in which the applicant for license renewal proposes an acceptable alternative for complying with specified portions of the regulations.

VI. BIBLIOGRAPHY

10 CFR Part 51. *Code of Federal Regulations*, Title 10, *Energy,* Part 51, "Environmental Protection Regulations for Domestic Licensing and Related Regulatory Functions."

U.S. Nuclear Regulatory Commission (NRC). 2013. *Generic Environmental Impact Statement for License Renewal of Nuclear Plants*. NUREG-1437, Vols. 1, 2, and 3, Revision 1. Office of Nuclear Reactor Regulation, Washington, D.C.

U.S. NUCLEAR REGULATORY COMMISSION
ENVIRONMENTAL STANDARD REVIEW PLAN
OFFICE OF NUCLEAR REACTOR REGULATION

2.0 ALTERNATIVES INCLUDING THE PROPOSED ACTION

I. AREAS OF REVIEW

This environmental standard review plan (ESRP) provides guidance for the preparation of the discussion of alternatives and the proposed action. The proposed action for license renewal is described in Section 2.1 of the *Generic Environmental Impact Statement for License Renewal of Nuclear Plants* (GEIS; NUREG-1437, Volumes 1, 2, and 3, Revision 1).

The purpose of this section is to (1) provide a statement of the proposed action (license renewal) for the SEIS and (2) provide background information related to the regulatory basis for license renewal and a brief description of the alternatives.

II. ACCEPTANCE CRITERIA

The reviewer should ensure that the introduction prepared under this ESRP is consistent with the following regulations:

- 10 CFR 51.70(b). The draft environmental impact statement will be concise, clear, and analytic, and written in plain language with appropriate graphics....The format provided in Section 1(a) of Appendix A of this subpart should be used. The Nuclear Regulatory Commission (NRC) staff will independently evaluate and be responsible for the reliability of all information used in the draft environmental impact statement.

- 10 CFR 51.95(c), concerning renewal of an operating license or combined license for a nuclear power plant. Under Parts 52 or 54 of this chapter, the Commission shall prepare an environmental impact statement, which is a supplement to the Commission's NUREG-1437, "Generic Environmental Impact Statement for License Renewal of Nuclear Plants."

- 10 CFR 51.103(a)(5). In making a final decision on a license renewal action pursuant to Part 54 of this chapter, the Commission shall determine whether or not the adverse environmental impacts of license renewal are so great that preserving the option of license renewal for energy planning decisionmakers would be unreasonable.

- 10 CFR Part 51, Appendix A to Subpart A of Part 51, concerning format for presentation of material in environmental impact statements

- 10 CFR Part 51, Appendix A(5), concerning alternatives including the proposed action

Technical Rationale

Renewal of a plant operating license is defined in 10 CFR Part 51 as a major Federal action requiring the preparation of an environmental impact statement (EIS). The introductory paragraphs prepared under this ESRP should clearly define the action and provide the readers of the supplemental environmental impact statement (SEIS) with background information related to license renewal. This information is summarized in the GEIS.

III. REVIEW PROCEDURES

The material to be prepared is informational in nature; no specific analysis of the data is required. Much of the required material may be taken directly from the GEIS. However, the reviewer should reflect the applicant's schedule for activities in preparation for license renewal, including refurbishment.

IV. EVALUATION FINDINGS

The reviewer for this ESRP should prepare several introductory paragraphs for the SEIS. The first paragraph should clearly state the nature of the proposed action (license renewal). The remaining paragraphs should describe the regulatory bases for license renewal, outline the process of license renewal, and outline the applicant's process.

V. IMPLEMENTATION

The method described in this ESRP would be used by the staff in evaluating conformance with the Commission's regulations, except in those cases in which the applicant for license renewal proposes an acceptable alternative for complying with specified portions of the regulations.

VI. BIBLIOGRAPHY

10 CFR Part 51. *Code of Federal Regulations*, Title 10, *Energy,* Part 51, "Environmental Protection Regulations for Domestic Licensing and Related Regulatory Functions."

U.S. Nuclear Regulatory Commission (NRC). 2013. *Generic Environmental Impact Statement for License Renewal of Nuclear Plants.* NUREG-1437, Vols. 1, 2, and 3, Revision 1. Washington, D.C.

U.S. NUCLEAR REGULATORY COMMISSION
ENVIRONMENTAL STANDARD REVIEW PLAN
OFFICE OF NUCLEAR REACTOR REGULATION

2.1 GENERAL PLANT INFORMATION

I. AREAS OF REVIEW

This environmental standard review plan (ESRP) provides guidance for the description of the plant and plant operations during the renewal term. This section includes a description of the layout and appearance of the nuclear plant facility and existing structures (onsite and offsite). It also includes descriptions of the reactor and electric generating equipment, as well as the plant's cooling system and auxiliary water systems.

The scope includes (1) description of principal structures, site boundaries, exclusion areas, restricted areas, and transportation routes to the site, (2) the type(s) and size(s) of reactors and electrical generating equipment and their major performance parameters, (3) a general description of the cooling system and modes of operation, (4) the intake and discharge locations and structures, (5) the auxiliary system, and (6) performance characteristics for these systems.

Data and Information Needs

The types of data and information needed are nuclear power plant site- and plant-specific. The following data or information may be needed, as appropriate:

- a map and description of the plant site location including State and local political jurisdictions (e.g., county, town, township, service districts, parish),

---

USNRC ENVIRONMENTAL STANDARD REVIEW PLAN

Environmental standard review plans are prepared for the guidance of the Office of Nuclear Reactor Regulation staff responsible for environmental reviews for nuclear power plants. These documents are made available to the public as part of the Commission's policy to inform the nuclear industry and the general public of regulatory procedures and policies. Environmental standard review plans are not substitutes for regulatory guides or the Commission's regulations and compliance with them is not required. These supplemental environmental standard review plans are keyed to Regulatory Guide 4.2, Supplement 1, "Preparation of Environmental Reports for Nuclear Power Plant License Renewal Applications."

Published environmental standard review plans will be revised periodically, as appropriate, to accommodate comments and to reflect new information and experience.

Comments and suggestions for improvement will be considered and should be sent to the U.S. Nuclear Regulatory Commission, Office of Nuclear Reactor Regulation, Washington, DC 20555-0001.

- maps of:
  - the site showing site boundaries and properties; plant exclusion area; site structures and facilities; major land uses (with land use classifications consistent with the U.S. Geological Survey categories; USGS 1997) and land cover; the construction zone for refurbishment, if any; sites for any other planned buildings, facilities, and structures (both temporary and permanent); areas under lease and public access; and transportation routes entering and adjacent to the site
  - the site vicinity within a 6-mile (10-kilometer) radius of the site showing boundaries of political jurisdictions, place names, topographic and physiographic features, residential areas, airports, industrial and commercial facilities, roads and highways, railroads, American Indian and/or Bureau of Indian Affairs lands held in trust for American Indians, Indian Tribes' lands, and military reservations
  - the region within a 50-mile (80-kilometer) radius of the site showing political jurisdictions, place names, topographic and physiographic features, and transportation networks and facilities

- Identification and description of known and reasonably foreseeable Federal and non-Federal projects and other actions that may contribute to the cumulative environmental impacts of license renewal and extended plant operation. Identify and map all Federal facilities, including national parks, national forests, national wildlife areas, and military facilities; American Indian and/or Bureau of Indian Affairs lands held in trust for American Indians; Indian Tribes' lands; and State parks, recreational areas, and conservation lands. Include distances, as well as nonattainment and maintenance areas defined under the Clean Air Act, as amended within 50 miles (80 kilometers) of the plant site.

- the number of units and description of each reactor, including type (e.g., boiling-water reactor and pressurized-water reactor), power conversion system manufacturer, fuel assembly description, and total quantities of uranium

- engineered safety features

- the historic average irradiation level of spent fuel, in megawatt days/ton

- the rated and design core thermal power, the rated and design gross electrical output, and the rated and design net electrical output in megawatts electric (MWe). (The rated power is defined as the power level at which each reactor is operated, and the design power is defined as the highest power level that would be permitted by plant design. The gross electrical output is the power level measured at the output terminals of the generator and expressed in MWe. The net unit electrical output is equal to the gross electrical output minus the nominal service and auxiliary loads.)

- a simplified flow diagram for the reactor-power conversion system

- a description of the plant's heat dissipation system, including the water supply source; intake and discharge locations; intake velocity; flow path of water from the intake point to the discharge point; any installed equipment or mitigation measures that reduce aquatic organism entrainment or impingement; and average temperatures of water at the discharge point. The description should include each operational mode and indicate the periods of time that the system has historically operated in each mode.

- for each operational mode:
    - the quantities of heat generated, dissipated to the atmosphere, and released in liquid discharges
    - identification of the water source and quantities of water withdrawn, consumed, and discharged

- monthly variation and stratification for the body of water used for cooling intake and discharge

- descriptions of changes to the cooling system in preparation for license renewal or during the renewal term

II. ACCEPTANCE CRITERIA

The reviewer should ensure that the introductory and descriptive paragraphs prepared under this ESRP are consistent with the following regulations:

- 10 CFR 51.52, concerning criteria related to plant-specific analysis of the effects of transportation of fuel and waste to and from the facility. Note: Generic determinations have been made that the impacts in Table S-4 are bounding for fuel with uranium enrichment of up to 5% by weight irradiated to 62,000 megawatt days per ton, provided that fuel is shipped more than 5 years after discharge from the reactor.

- 10 CFR 51.53(c)(2). The report must contain a description of the proposed action, including the applicant's plans to modify the facility or its administrative control procedures as described in accordance with 10 CFR 54.21 of this chapter. This report must describe in detail the affected environment around the plant, the modifications directly affecting the environment or any plant effluents, and any planned refurbishment activities. In addition, the applicant shall discuss in this report the environmental impacts of alternatives and any other matters discussed in 10 CFR 51.45.

- 10 CFR 51.53(c)(3)(ii)(A-D) describe the analyses which must be performed with respect to the environmental impacts of and related interactions with the environment of a plant's cooling water and auxiliary systems and which necessitates that the environmental report provide a description of such systems, including their water requirements and intakes and discharges, to support the discussions of the affected environment.

- 10 CFR 51.70(b). The draft environmental impact statement will be concise, clear, and analytic, and written in plain language with appropriate graphics....The format provided in Section 1(a) of Appendix A of this subpart should be used. The Nuclear Regulatory Commission (NRC) staff will independently evaluate and be responsible for the reliability of all information used in the draft environmental impact statement.

Technical Rationale

The technical rationale for evaluating the applicant's external appearance and setting description is discussed in the following paragraph:

> A description of the overall appearance of the nuclear power plant and its setting is needed to clarify the physical parameters of the current power station and any significant modifications to the facility. The description of the external appearance of the plant and plant layout should be in sufficient detail to form an adequate basis for staff analysis of various land-use and socioeconomic impacts of continued plant operations and refurbishment.

The technical rationale for evaluating the description of the applicant's reactor system is discussed in the following paragraph:

> A description of the overall nuclear energy generating system is useful background information for the evaluation of certain environmental impacts resulting from continued plant operations and refurbishment activities. This description should include information about reactor type, number of units, thermal power level, and other factors about the facility.

The technical rationale for evaluating the description of the applicant's cooling systems is discussed in the following paragraph:

> The cooling system has the greatest effect on the environment. This section is descriptive in nature and presents information necessary for the evaluation of environmental impacts associated with cooling system modification related to license renewal and continued plant operations during the renewal term. The description of the external appearance of the cooling system and its operational modes should be in sufficient detail to form an adequate basis for staff analysis of the environmental impacts of continued plant operations and refurbishment activities during the license renewal term.

III. REVIEW PROCEDURES

The reviewer should ensure that the description of the layout and appearance of the nuclear plant facility and existing structures (onsite and offsite) provides adequate information for the reviews conducted under the ESRPs in Chapters 3 and 4. The following review steps are suggested:

1. Review plant and station layout and external appearance data.

2. Determine the relationship of the plant design and layout to the surrounding environment, including any aesthetic features of the site and vicinity.

3. Identify maps and drawings that show relevant features of the plant, the site, and the region. The maps and drawings should also identify significant offsite features, if any, in the vicinity (i.e., Federal facilities, including national parks, forests, wildlife areas, Indian and/or Bureau of Indian Affairs lands held in trust for Indians, and Indian Tribes' lands).

The material to be prepared on the reactor-power conversion system is informational in nature; no specific analysis of the data is required. Identify the reactor power conversion and engineered safety feature systems and the basic design performance data. As a rule, if the data listed under "Data and Information Needs" above are provided, that objective would be met.

The material to be prepared on the cooling systems is informational in nature. No specific analysis is required, but the use of tables such as Table 2.1.3-1 and Table 2.1.3-2 in this ESRP may assist data organization. The reviewer should gather the following information largely from design and historical documentation for use in later sections:

- For the general cooling system description, include:

  - type and configuration

  - water source and proximity to facility

  - modes of operation and percentage of time, water source and quantities of water withdrawn, consumed, and discharged in each mode

  - specific details depending on system type (see Tables 2.1.3-1 and 2.1.3-2)

  - monthly variation and stratification for the body of water used for cooling intake and discharge

  - other major plant systems and flow rates

- For intake systems include:

  - a drawing of the intake structure showing the relationship of the structure to the water surface, bottom geometry, and shoreline

  - the location of the intake in relation to the outfall

  - a description of cooling-water pumping facility

  - a description of the trash racks, traveling screens, trash baskets, and fish return devices

  - performance characteristics (e.g., flow rates, intake velocities) for the operational modes identified

  - performance characteristics for specific intake related functions, such as de-icing, trash rack clearing, screen washing, trash basket removal, or fish return system operation

  - the location and description of components for the addition of chemicals (e.g., corrosion inhibitors, antifouling agents) to the intake system.

- For discharge systems include:

  - drawings of the outfall structure, showing its location in the receiving water body, relationship to water surface, bottom geometry, and shoreline

  - a description of discharge canal or discharge lines

  - performance characteristics (e.g., discharge flow rates, discharge velocities, discharge temperatures, and temperature differentials) for the operational modes identified

  - descriptions of specific discharge related components (e.g., diffusers, fish barriers)

- For heat-dissipation systems include:

  - the location of heat-dissipation system components relative to other site features

  - the design details of heat-dissipation system components affecting system performance

  - heat-dissipation system performance characteristics for the operational modes

  - nuclear power plant site-specific meteorological data

  - nuclear power plant site-specific water supply data

- For cooling towers, determine average discharge temperatures for each month of the year using cooling tower performance curves. The average discharge temperature would be calculated by using the average wet-bulb temperature for the month.

- For spray systems, analyze the applicant's estimates of average monthly discharge temperatures. The depth and extent of this analysis should depend on the seriousness of the predicted impacts of the heated effluent on the receiving body of water and the level of confidence in the applicant's model.

- In the cases where auxiliary systems are employed to further cool the blowdown discharged from the main cooling system, determine the final discharge temperature.

Table 2.1.3-1. Design Details of Heat-Dissipation-System Components

Component	Design Details
Cooling towers (from the ER)	Type Configuration Materials of construction Number and arrangement Rated heat-dissipation capacity
Cooling lakes and ponds (from the ER)	Surface area Volume Bathymetry
Spray ponds or canals (from the ER)	Arrangement and configuration of spray modules Pond or canal geometry Surface area and water volume
Condenser (from the ER)	Heat transfer area and materials of construction Antifouling treatment

Table 2.1.3-2. Performance Characteristics of the Heat-Dissipation System

Component	Design Details
Cooling towers (from the ER)	Input and discharge flow rates and temperatures for monthly average meteorological conditions Wet-bulb temperature, approach to wet-bulb, and range Performance curves Air flow Power consumption Noise levels Drift rate and drop size
Cooling lakes and ponds (from the ER)	Flow rates (through condenser) Flow-through times Flow pattern Monthly average water temperatures (mean for entire lake or pond, inlet [from condenser], outlet [to condenser]) Surface elevation (mean, maximum, minimum)
Spray ponds or canals (from the ER)	Flow rates (through condenser) Flow-through times Flow pattern Monthly average water temperatures (inlet [from condenser], outlet [to condenser]) Surface elevation (mean, maximum, minimum) Spray system operating parameters (e.g., power consumption, drop size)
Once-through systems (from the ER)	Condenser flow rate Temperature differential across condenser Time-of-passage through system (including intake and discharge system passage times)

IV. EVALUATION FINDINGS

The depth and extent of the input to the SEIS should be governed by land-use considerations that could be affected by the layout of the nuclear power plant and supporting structures. It should include a summary description of the reactor-power conversion and engineered safety feature systems, a flow diagram, and a table of design and performance parameters.

The level of detail of information included in the SEIS should include the following information:

- narrative description of the cooling system and the intake and discharge structures and characteristics

- sketches of intake, discharge, and heat-dissipation components

- description of operational modes and their important characteristics (e.g., frequency and duration, discharge temperature, water consumption, and chemical concentration factor)

- drawings of important subsystems (e.g., perforated-pipe assemblies)

The reviewer should verify that cooling system component descriptions are consistent, accurate, and given in sufficient detail to serve the needs of the reviewers of intake, discharge, and heat-dissipation system impacts.

V. IMPLEMENTATION

The method described in this ESRP would be used by the staff in evaluating conformance with the Commission's regulations, except in those cases in which the applicant for license renewal proposes an acceptable alternative for complying with specified portions of the regulations.

VI. BIBLIOGRAPHY

10 CFR Part 51. *Code of Federal Regulations*, Title 10, *Energy*, Part 51, "Environmental Protection Regulations for Domestic Licensing and Related Regulatory Functions."

Clean Air Act of 1970, as amended. 42 USC 7401 et seq.

U.S. Geological Survey (USGS). 1997. *USGS Land Use and Land Cover Data*. USGS Survey Earth Resources Observation Data Center, Sioux Falls, South Dakota.

U.S. Nuclear Regulatory Commission (NRC). 2013. *Generic Environmental Impact Statement for License Renewal of Nuclear Plants*. NUREG-1437, Vols. 1, 2, and 3, Revision 1. Office of Nuclear Reactor Regulation, Washington, D.C.

U.S. NUCLEAR REGULATORY COMMISSION
ENVIRONMENTAL STANDARD REVIEW PLAN
OFFICE OF NUCLEAR REACTOR REGULATION

2.2 REFURBISHMENT ACTIVITIES

I. AREAS OF REVIEW

This environmental standard review plan (ESRP) provides guidance for the description of any planned refurbishment activities performed in support of license renewal. This section includes a description of any major structures and components that would be replaced or modified.

The scope includes (1) identification and description of major structures and components to undergo refurbishment, (2) where construction materials would be stored, as well as removal and disposal, (3) description of related activities that have the potential to affect the surrounding environment either directly or indirectly.

Data and Information Needs

The types of data and information needed would be affected by nuclear power plant site- and plant-specific factors. The following data or information may be needed:

- description of the proposed refurbishment activity, including specific structures and components that would be replaced or modified

- description of the location where materials would be stored, as well as removal and disposal

- description of any transportation or delivery activities in support of the refurbishment activity, including the transport and delivery of equipment, structures, and components (e.g., steam generators, vessel heads), as well as any dredging and bridge and road modifications

- list of applicable Federal and State agency permits required for this activity

- description of specific refurbishment-related activities that have the potential to either directly or indirectly affect the environment

- discussion of the schedule for the refurbishment activity, including normal maintenance schedules and refueling outages

II. ACCEPTANCE CRITERIA

The reviewer should ensure that the introductory and descriptive paragraphs prepared under this ESRP are consistent with the following regulations:

- 10 CFR 51.53(c)(2). The report must contain a description of the proposed action, including the applicant's plans to modify the facility or its administrative control procedures as described in accordance with 10 CFR 54.21 of this chapter. This report must describe in detail the affected environment around the plant, the modifications directly affecting the environment or any plant effluents, and any planned refurbishment activities. In addition, the applicant shall discuss in this report the environmental impacts of alternatives and any other matters discussed in 10 CFR 51.45.

- 10 CFR 51.70(b). The draft environmental impact statement will be concise, clear, and analytic, and written in plain language with appropriate graphics....The format provided in Section 1(a) of Appendix A of this subpart should be used. The Nuclear Regulatory Commission (NRC) staff will independently evaluate and be responsible for the reliability of all information used in the draft environmental impact statement.

Technical Rationale

This section is descriptive in nature and presents information necessary for the evaluation of environmental impacts associated with refurbishment. The descriptions should be in sufficient detail to form an adequate basis for staff analysis of environmental impacts of refurbishment activities associated with license renewal.

III. REVIEW PROCEDURES

The material to be prepared on refurbishment activities is informational; no specific analysis is required. The reviewer should ensure that description of the plant refurbishment activities provides adequate information for the reviews conducted under the ESRPs in Chapters 3 and 4. The following review steps are suggested:

1. Review the discussion of plant refurbishment in the *Generic Environmental Impact Statement for License Renewal of Nuclear Power Plant* (GEIS; NUREG-1437).

2. Obtain a description of the proposed refurbishment activity, including descriptions of the specific structures and components that would be replaced or modified.

3. Obtain descriptions of transport and storage of necessary equipment and materials, including any proposed transportation plans.

4. Obtain the proposed schedule for refurbishment work, including planned changes in staffing, if any.

5. Prepare a section describing the refurbishment activities for the supplemental environmental impact statement (SEIS).

IV. EVALUATION FINDINGS

The reviewer of information covered by this ESRP should prepare introductory paragraphs for the SEIS. The paragraph(s) should introduce the nature of the material to be presented.

V. IMPLEMENTATION

The method described in this ESRP would be used by the staff in evaluating conformance with the Commission's regulations, except in those cases in which the applicant for license renewal proposes an acceptable alternative for complying with specified portions of the regulations.

VI. BIBLIOGRAPHY

10 CFR Part 51. *Code of Federal Regulations*, Title 10, *Energy*, Part 51, "Environmental Protection Regulations for Domestic Licensing and Related Regulatory Functions."

U.S. Nuclear Regulatory Commission (NRC). 2013. *Generic Environmental Impact Statement for License Renewal of Nuclear Plants*. NUREG-1437, Vols. 1, 2, and 3, Revision 1. Office of Nuclear Reactor Regulation, Washington, D.C.

U.S. NUCLEAR REGULATORY COMMISSION
ENVIRONMENTAL STANDARD REVIEW PLAN
OFFICE OF NUCLEAR REACTOR REGULATION

2.3 EMPLOYMENT

I. AREAS OF REVIEW

This environmental standard review plan (ESRP) provides guidance for the discussion of employment at nuclear power plants. The discussion should include both permanent full-time onsite and refueling outage employment (i.e., the total number of full-time applicant and contractor employees), as well as information on the average duration of refueling and maintenance outages (number of weeks) and their frequency (number of months).

The reviewer should evaluate the applicant's projections of the increase in the number of workers required to support any refurbishment activity described in Section 2.2. The reviewer should also estimate the residential distribution of the refurbishment workforce.

Data and Information Needs

The types of data and information needed would be affected by nuclear power plant site- and plant-specific factors. The following data or information may be needed:

- a description of the plant's operational employment history, including maintenance and refueling outages

- plant staffing history, including maximum annual average number of workers supporting plant operations, maintenance, and refueling outages

- anticipated number of workers onsite during refurbishment activities

- anticipated changes in the number of workers onsite during and in support of license renewal

II. ACCEPTANCE CRITERIA

The reviewer should ensure that the introductory and descriptive paragraphs prepared under this ESRP are consistent with the following regulation:

- 10 CFR 51.53(c)(2). The report must contain a description of the proposed action, including the applicant's plans to modify the facility or its administrative control procedures as described in accordance with 10 CFR 54.21 of this chapter. This report must describe in detail the affected environment around the plant, the modifications directly affecting the environment or any plant effluents, and any planned refurbishment activities. In addition, the applicant shall discuss in this report the environmental impacts of alternatives and any other matters discussed in 10 CFR 51.45.

- 10 CFR 51.70(b). The draft environmental impact statement will be concise, clear, and analytic, and written in plain language with appropriate graphics....The format provided in Section 1(a) of Appendix A of this subpart should be used. The NRC staff will independently evaluate and be responsible for the reliability of all information used in the draft environmental impact statement.

Technical Rationale

This section is descriptive in nature and presents information necessary for the evaluation of environmental impacts associated with plant operations. The descriptions should be in sufficient detail to form an adequate basis for analysis of the environmental impacts of continued plant operations during the license renewal term and refurbishment.

III. REVIEW PROCEDURES

The material to be prepared on plant employment is informational in nature. No specific analysis is required. The following review steps are suggested:

1. Review the discussion of plant employment in the *Generic Environmental Impact Statement for License Renewal of Nuclear Power Plant* (GEIS; NUREG-1437).

2. Obtain a description of the plant employment history, including maintenance and refueling outages and annual average employment.

3. Obtain anticipated changes in the number of workers onsite during and in support of license renewal and refurbishment, if any.

IV. EVALUATION FINDINGS

The depth and extent of the input to the supplemental environmental impact statement (SEIS) would depend on nuclear power plant site- and plant-specific factors. The reviewer should verify that the plant employment description is consistent, accurate, and given in sufficient detail to serve the needs of the reviewers for ESRP sections in Chapters 3 and 4.

V. IMPLEMENTATION

The method described in this ESRP would be used by the staff in evaluating conformance with the Commission's regulations, except in those cases in which the applicant for license renewal proposes an acceptable alternative for complying with specified portions of the regulations.

VI. BIBLIOGRAPHY

10 CFR Part 51. *Code of Federal Regulations*, Title 10, *Energy,* Part 51, "Environmental Protection Regulations for Domestic Licensing and Related Regulatory Functions."

U.S. Nuclear Regulatory Commission (NRC). 2013. *Generic Environmental Impact Statement for License Renewal of Nuclear Plants*. NUREG-1437, Vols. 1, 2, and 3, Revision 1. Office of Nuclear Reactor Regulation, Washington, D.C.

2.4 DESCRIPTION OF THE ALTERNATIVES TO THE PROPOSED ACTION

I. AREAS OF REVIEW

This environmental standard review plan (ESRP) provides guidance for describing how the applicant identified and selected reasonable alternatives to the proposed action.

The scope includes (1) a brief description the process used to identify and select alternatives to the proposed action by the applicant, and (2) brief descriptions of all of the replacement power alternatives considered by the applicant.

Data and Information Needs

The types of data and information needed would be affected by nuclear power plant site- and plant-specific factors. The following data or information may be needed:

- description of the process used by the applicant to determine reasonable replacement power alternatives to the proposed action

- description of all alternatives considered

- indication of which alternatives should be described in further detail

- brief description of any alternatives considered that would reduce or avoid adverse effects

---

USNRC ENVIRONMENTAL STANDARD REVIEW PLAN

II. ACCEPTANCE CRITERIA

The reviewer should ensure that the paragraphs prepared under this ESRP are consistent with the following regulations:

- 10 CFR 51.45(b)(3), concerning alternatives to the proposed action. The discussion of alternatives shall be sufficiently complete to aid the Commission in developing and exploring, pursuant to Section 102(2)(E) of the National Environmental Policy Act of 1069 (NEPA), "appropriate alternatives to recommended courses of action in any proposal which involves unresolved conflicts concerning alternative uses of available resources." To the extent practicable, the environmental impacts of license renewal and the replacement power alternatives should be presented in comparative form.

- 10 CFR 51.53(c)(2). The report must contain a description of the proposed action, including the applicant's plans to modify the facility or its administrative control procedures as described in accordance with 10 CFR 54.21 of this chapter. This report must describe in detail the affected environment around the plant, the modifications directly affecting the environment or any plant effluents, and any planned refurbishment activities. In addition, the applicant shall discuss in this report the environmental impacts of alternatives and any other matters discussed in 10 CFR 51.45.

- 10 CFR 51.53(c)(3)(iii). The report must contain a consideration of alternatives for reducing adverse impacts, as required by Section 51.45(c), for all Category 2 license renewal issues in Appendix B to Subpart A of this part. No such consideration is required for Category 1 issues in Appendix B to Subpart A of this part.

- 10 CFR 51.70(b). The draft environmental impact statement will be concise, clear, and analytic, and written in plain language with appropriate graphics.…The format provided in Section 1(a) of Appendix A of this subpart should be used. The Nuclear Regulatory Commission (NRC) staff will independently evaluate and be responsible for the reliability of all information used in the draft environmental impact statement.

- 10 CFR 51.103(a)(5). In making a final decision on a license renewal action pursuant to Part 54 of this chapter, the Commission shall determine whether or not the adverse environmental impacts of license renewal are so great that preserving the option of license renewal for energy planning decisionmakers would be unreasonable.

- 10 CFR Part 51, Appendix A to Subpart A of Part 51, concerning format for presentation of material in environmental impact statements

- 10 CFR Part 51, Appendix A to Subpart A of Part 51(5), concerning alternatives including the proposed action

Technical Rationale

The GEIS does not contain any conclusions regarding the environmental impact or acceptability of alternatives to license renewal. Accordingly, the NRC must conduct an analysis of reasonable alternatives to license renewal in plant-specific environmental reviews. A reasonable alternative must be commercially viable on a utility scale and operational prior to the expiration of the reactor's operating license, or expected to become commercially viable on a utility scale and operational prior to the expiration of the reactor's operating license. Analysis of replacement power alternatives does not involve the determination of whether any power is needed or should be generated. The decision to generate power and the determination of how much power is needed are at the discretion of State and utility officials. The potential environmental impacts evaluated include land use, ecology, aesthetics, water quality, air quality, solid waste, human health, socioeconomics, environmental justice, and historic and cultural resources.

III. REVIEW PROCEDURES

The material to be prepared for the plant description of alternatives to the proposed action is informational in nature; no specific analysis is required. The following review steps are suggested:

1. Obtain a description of the process used by the applicant to determine reasonable replacement power alternatives to the proposed action.

2. Obtain a list of alternatives considered by the applicant and state authorities.

3. Obtain a list of which of the considered alternatives were selected to be described in greater detail.

4. Prepare a section describing the considered alternatives to the proposed action for the supplemental environmental impact statement (SEIS).

IV. EVALUATION FINDINGS

The information in the SEIS would depend on nuclear power plant site- and plant-specific factors.

V. IMPLEMENTATION

The method described in this ESRP would be used by the staff in evaluating conformance with the Commission's regulations, except in those cases in which the applicant for license renewal proposes an acceptable alternative for complying with specified portions of the regulations.

VI. BIBLIOGRAPHY

10 CFR Part 51. *Code of Federal Regulations*, Title 10, *Energy,* Part 51, "Environmental Protection Regulations for Domestic Licensing and Related Regulatory Functions."

National Environmental Policy Act of 1969, as amended (NEPA). 42 USC 4321 et seq.

U.S. Nuclear Regulatory Commission (NRC). 2013. *Generic Environmental Impact Statement for License Renewal of Nuclear Plants* (GEIS). NUREG-1437, Vols. 1, 2, and 3, Revision 1. Office of Nuclear Reactor Regulation, Washington, D.C.

U.S. NUCLEAR REGULATORY COMMISSION
ENVIRONMENTAL STANDARD REVIEW PLAN
OFFICE OF NUCLEAR REACTOR REGULATION

3.0 AFFECTED ENVIRONMENT

I. AREAS OF REVIEW

This environmental standard review plan (ESRP) provides guidance for preparing the introduction to sections that describe the plant's affected environment based on the reviews conducted under ESRP Sections 3.1 through 3.11.

The scope includes (1) review of the affected environment in the *Generic Environmental Impact Statement for License Renewal of Nuclear Plants* (GEIS; NUREG-1437, Volumes 1, 2, and 3, Revision 1) and (2) preparation of input to the supplemental environmental impact statement (SEIS).

II. ACCEPTANCE CRITERIA

The reviewer should ensure that the introductory paragraphs prepared under this ESRP are consistent with the following regulations:

*   10 CFR 51.45(d), concerning status of compliance. The environmental report shall list all Federal permits, licenses, approvals and other entitlements which must be obtained in connection with the proposed action and shall describe the status of compliance with these requirements. The environmental report shall also include a discussion of the status of compliance with applicable environmental quality standards and requirements including, but not limited to, applicable zoning and land-use regulations, and thermal and other water pollution limitations or requirements which have been imposed by Federal, State, regional, and local agencies having responsibility for environmental protection.

3.0-1

- 10 CFR 51.53(c)(2). The report must contain a description of the proposed action, including the applicant's plans to modify the facility or its administrative control procedures as described in accordance with 10 CFR 54.21 of this chapter. This report must describe in detail the affected environment around the plant, the modifications directly affecting the environment or any plant effluents, and any planned refurbishment activities. In addition, the applicant shall discuss in this report the environmental impacts of alternatives and any other matters discussed in 10 CFR 51.45.

- 10 CFR 51.70(b). The draft environmental impact statement will be concise, clear, and, analytic, and written in plain language with appropriate graphics....The format provided in Section 1(a) of Appendix A of this subpart should be used. The Nuclear Regulatory Commission (NRC) staff will independently evaluate and be responsible for the reliability of all information used in the draft environmental impact statement.

- 10 CFR 51.95(c), concerning renewal of an operating license or combined license for a nuclear power plant. Under Parts 52 or 54 of this chapter, the Commission shall prepare an environmental impact statement, which is a supplement to the Commission's NUREG-1437, "Generic Environmental Impact Statement for License Renewal of Nuclear Plants."

- 10 CFR Part 51, Appendix A to Subpart A, para. 6, concerning affected environment. The environmental impact statement will succinctly describe the environment to be affected by the proposed action. Data and analyses in the statement will be commensurate with the importance of the impact, with less important material summarized, consolidated, or simply referenced. Effort and attention will be concentrated on important issues; useless bulk will be eliminated.

- 10 CFR Part 51, Appendix B to Subpart A, "Environmental Effect of Renewing the Operating License of a Nuclear Power Plant," Table B-1, "Summary of Findings on NEPA Issues for License Renewal of Nuclear Power Plants"

Technical Rationale

The review conducted under this ESRP leads to preparation of a section describing the affected environment for the SEIS that provides background information to be used in evaluating environmental impacts of continued plant operations and refurbishment activities associated with license renewal.

III. REVIEW PROCEDURES

The material to be prepared is informational and descriptive; no specific analysis of data is required. The introduction should list the information to be presented and describe its relationship to the environmental consequences to be presented in Chapter 4 of the SEIS. It should indicate that the objective of SEIS Subsections 3.1 through 3.11 is to provide a general description of the affected environment as background and/or baseline information. Some detailed descriptions may be needed to support the analyses of environmental impacts in Chapter 4.

It is important to point out specific sections of this chapter that address environmental issues raised by the public in scoping meetings or in correspondence on the license renewal application.

IV. EVALUATION FINDINGS

The reviewer for this ESRP should prepare a paragraph(s) to introduce the nature of the material to be presented by the reviewers of information covered by ESRPs 3.1 through 3.11. The depth and extent of the input to the SEIS would be governed, in part, by the extent of the potential impacts of continued operations during the license renewal term and refurbishment activities in support of license renewal.

V. IMPLEMENTATION

The method described in this ESRP would be used by the staff in evaluating conformance with the Commission's regulations, except in those cases in which the applicant for license renewal proposes an acceptable alternative for complying with specified portions of the regulations.

VI. BIBLIOGRAPHY

10 CFR Part 51. *Code of Federal Regulations*, Title 10, *Energy,* Part 51, "Environmental Protection Regulations for Domestic Licensing and Related Regulatory Functions."

National Environmental Policy Act of 1969, as amended (NEPA). 42 USC 4321 et seq.

U.S. Nuclear Regulatory Commission (NRC). 2013. *Generic Environmental Impact Statement for License Renewal of Nuclear Plants*. NUREG-1437, Vols. 1, 2, and 3, Revision 1. Office of Nuclear Reactor Regulation, Washington, D.C.

U.S. NUCLEAR REGULATORY COMMISSION

ENVIRONMENTAL STANDARD REVIEW PLAN

OFFICE OF NUCLEAR REACTOR REGULATION

3.1 LAND USE AND VISUAL RESOURCES

I. AREAS OF REVIEW

This environmental standard review plan (ESRP) provides guidance for the review of land use and visual resources of the site and vicinity, the region, and the in-scope transmission lines and corridors.

The scope should include establishing the nature and extent of present and planned land use and visual impacts within areas that might be impacted by continued plant operations during the license renewal term or modified as a result of refurbishment in support of license renewal.

Data and Information Needs

The types of data and information needed would be affected by nuclear power plant site- and plant-specific factors and presented in conjunction with the maps included in Section 2.1. The degree of detail should be scaled according to the anticipated magnitude of the potential impact. The following data or information may be needed, if appropriate:

- land-use data (onsite and offsite) and descriptions from prior environmental documents, including the applicant's environmental report (ER) and final environmental statements (FESs) prepared for nuclear plant construction and operation

- description of the plant site location in relationship to State and local political jurisdictions (e.g., county, town, township, service districts, parish)

- description of the site boundaries and properties; plant exclusion area; site structures and facilities; major land uses and land cover; the construction zone for refurbishment, if any; sites for any other planned buildings, facilities, and structures (both temporary and permanent); areas under lease or with public access; and transportation routes entering and adjacent to the site

- description of the site vicinity within a 6-mile (10-kilometer) radius regarding political jurisdictions, major land uses and land cover, topographic and physiographic features, transportation networks and facilities, place names, Indian and/or Bureau of Indian Affairs lands held in trust for Indians, and Indian Tribes' lands

- description of the region within a 50-mile (80-kilometer) radius of the site regarding political jurisdictions, place names, topographic and physiographic features, and transportation networks and facilities

- identification and description of known and reasonably foreseeable Federal and non-Federal projects and other actions that may contribute to the cumulative environmental impacts of license renewal and extended plant operation: all Federal facilities, including national parks, national forests, national wildlife areas, and military facilities; American Indian and/or Bureau of Indian Affairs lands held in trust for Indians; Indian Tribes' lands; and State parks, recreational areas, and conservation lands. Include distances, as well as nonattainment and maintenance areas defined under the Clean Air Act, as amended, within 50 miles (80 kilometers) of the plant site

- description of land uses and land cover within the in-scope transmission lines and corridors and any current and planned restrictions or covenants on use

- description of the plant's visual setting, including the identities and heights of the tallest visible plant structures, significant lights and vapor plumes, as well as direction and distances from which these structures, lights, and plumes are visible

- description of the existing visual impacts of in-scope transmission lines and corridors' level of significance, distance, and type of observer

II. ACCEPTANCE CRITERIA

Acceptance criteria for the review of onsite and offsite land use and the transmission lines and corridors are based on the following regulations:

- 10 CFR 51.45(d), concerning status of compliance. The environmental report shall list all Federal permits, licenses, approvals, and other entitlements which must be obtained in connection with the proposed action and shall describe the status of compliance with these requirements. The environmental report shall also include a discussion of the status of compliance with applicable environmental quality standards and requirements including, but not limited to, applicable zoning and land-use regulations, and thermal and other water pollution limitations or requirements which have been imposed by Federal, State, regional, and local agencies having responsibility for environmental protection.

- 10 CFR 51.53(c)(2). The report must contain a description of the proposed action, including the applicant's plans to modify the facility or its administrative control procedures as described in accordance with 10 CFR 54.21 of this chapter. This report must describe in detail the affected environment around the plant, the modifications directly affecting the environment or any plant effluents, and any planned refurbishment activities. In addition, the applicant shall discuss in this report the environmental impacts of alternatives and any other matters discussed in 10 CFR 51.45.

- 10 CFR 51.70(b). The draft environmental impact statement will be concise, clear, and analytic, and written in plain language with appropriate graphics....The format provided in Section 1(a) of Appendix A of this subpart should be used. The Nuclear Regulatory Commission (NRC) staff will independently evaluate and be responsible for the reliability of all information used in the draft environmental impact statement.

- 10 CFR Part 51, Appendix A to Subpart A, para. 6, concerning affected environment. The environmental impact statement will succinctly describe the environment to be affected by the proposed action. Data and analyses in the statement will be commensurate with the importance of the impact, with less important material summarized, consolidated, or simply referenced. Effort and attention will be concentrated on important issues; useless bulk will be eliminated.

- 15 CFR Part 930, the regulations governing implementation of the requirement for Federal consistency with approved coastal management programs (as set forth in the Coastal Zone Management Act of 1972).

Technical Rationale

The review conducted under this ESRP leads to preparation of a section describing the affected environment for the supplemental environmental impact statement (SEIS) that provides background information to be used in evaluating land-use impacts of continued plant operations and refurbishment activities associated with license renewal.

III. REVIEW PROCEDURES

The reviewer's analysis of land-use characteristics should be closely linked with the impact assessment review described in ESRP Section 4.1 to establish the land-use characteristics most likely to be affected by license renewal. The following review steps are suggested:

1. Review the applicant's ER, scoping issues, and any other applicable land-use information.

2. Identify onsite and offsite land uses and land cover, including areas within the in-scope transmission lines and corridors that could be affected by continued plant operations and refurbishment associated with license renewal.

3. Describe political jurisdictions, place names, topographic and physiographic features, and transportation networks and facilities.

4. Identify any current and planned site and local land-use, zoning, and development plans relevant to population and housing growth and control and changes in land use patterns.

5. Identify known and reasonably foreseeable Federal and non-Federal projects and other actions within the region that may contribute to the cumulative environmental impacts of license renewal and extended plant operation.

6. Identify potential impacts to a coastal zone or coastal watershed, as defined by each State participating in the National Coastal Zone Management Program.

7. Identify any visual impacts of the plant and of the in-scope transmission lines from various viewing locations.

IV. EVALUATION FINDINGS

The reviewer should ensure that the land-use information is adequate as a basis for assessment of the effects of continued plant operations and refurbishment activities associated with license renewal. The reviewer should consult with appropriate Federal, State, regional, local, and affected American Indian Tribes to assess the accuracy of the land-use designations, if necessary.

V. IMPLEMENTATION

The method described in this ESRP would be used by the staff in evaluating conformance with the Commission's regulations, except in those cases in which the applicant for license renewal proposes an acceptable alternative for complying with specified portions of the regulations.

VI. BIBLIOGRAPHY

10 CFR Part 51. *Code of Federal Regulations*, Title 10, *Energy,* Part 51, "Environmental Protection Regulations for Domestic Licensing and Related Regulatory Functions."

15 CFR Part 930. *Code of Federal Regulations*, Title 15, *Commerce and Foreign Trade*, Part 930, "Federal Consistency with Approved Coastal Management Programs."

Clean Air Act of 1970, as amended. 42 USC 7401 et seq.

Coastal Zone Management Act of 1972 (CZMA), as amended. 16 USC 1451 et seq.

U.S. Nuclear Regulatory Commission (NRC). 2013. *Generic Environmental Impact Statement for License Renewal of Nuclear Plants.* NUREG-1437, Vols. 1, 2, and 3, Revision 1. Office of Nuclear Reactor Regulation, Washington, D.C.

U.S. NUCLEAR REGULATORY COMMISSION
ENVIRONMENTAL STANDARD REVIEW PLAN
OFFICE OF NUCLEAR REACTOR REGULATION

3.2 METEOROLOGY, AIR QUALITY, AND NOISE

I. AREAS OF REVIEW

This environmental standard review plan (ESRP) provides guidance for the review of the meteorology, air quality, and noise environment of the site and surrounding area, and characterization of atmospheric transport. This review should provide background information for inclusion in the supplemental environmental impact statement (SEIS) and input to reviewers for ESRPs for license renewal dealing with evaluation of the impacts of continued plant operations during the license renewal term and refurbishment activities in support of license renewal.

The scope includes descriptions of (1) regional climatology, (2) meteorological characteristics of the site and vicinity using data from the onsite meteorological monitoring program, (3) local and regional atmospheric transport and diffusion characteristics, (4) local and regional air quality, and (5) noise generation of, and in the vicinity of, the site.

Data and Information Needs

The types of data and information needed would be affected by nuclear power plant site- and plant-specific factors; the level of detail should be scaled according to the anticipated magnitude of the potential impacts. The following data or information may be needed, if appropriate:

- climatic descriptions from prior environmental documents, including the environmental impact statements (EISs) prepared at the construction-permit and operating-license stages

---

- recent climatological data from nearby National Weather Service (NWS) stations

- extreme weather events, such as floods, hails, thunderstorms, tornadoes, hurricanes, etc., from the National Climatic Data Center Storm Events database, and historical events and damages to the site or nearby areas

- summary of meteorological data from the onsite meteorological program for the most recent 5-year period

- descriptions of meteorological phenomena, if any, associated with the plant's cooling system operation

- a description of regional air quality, including the locations of mandatory Federal Class I areas

- topographic data if substantially different from the data presented in earlier EISs

- lists of nonattainment and/or maintenance areas in the region

- a map of the region within a 50-mile (80-kilometer) radius of the nonattainment and maintenance areas of the site

- any current or past noise studies and analyses conducted in the vicinity of the site

- nearby sensitive receptors such as residences, schools, nursing homes, biota, and habitat

- list of loudest-noise-generating facilities and activities in the surrounding area, with given distances to the nearest site boundary

II. ACCEPTANCE CRITERIA

Acceptance criteria for the evaluation of site meteorology, air quality, and noise are based on the following regulations:

- 10 CFR 51.45(d), "Status of Compliance." The environmental report shall list all Federal permits, licenses, approvals, and other entitlements which must be obtained in connection with the proposed action and shall describe the status of compliance with these requirements. The environmental report shall also include a discussion of the status of compliance with applicable environmental quality standards and requirements including, but not limited to, applicable zoning and land-use regulations, and thermal and other water pollution limitations or requirements which have been imposed by Federal, State, regional, and local agencies having responsibility for environmental protection.

- 10 CFR 51.53(c)(2). The report must contain a description of the proposed action, including the applicant's plans to modify the facility or its administrative control procedures as described in accordance with 10 CFR 54.21 of this chapter. This report must describe in detail the affected environment around the plant, the modifications directly affecting the environment or any plant effluents, and any planned refurbishment activities. In addition, the applicant shall discuss in this report the environmental impacts of alternatives and any other matters discussed in 10 CFR 51.45.

- 10 CFR 51.70(b). The draft environmental impact statement will be concise, clear, and analytic, and written in plain language with appropriate graphics....The format provided in Section 1(a) of Appendix A of this subpart should be used. The NRC staff will independently evaluate and be responsible for the reliability of all information used in the draft environmental impact statement.

- 10 CFR Part 51, Appendix A to Subpart A, para. 6, concerning affected environment. The environmental impact statement will succinctly describe the environment to be affected by the proposed action. Data and analyses in the statement will be commensurate with the importance of the impact, with less important material summarized, consolidated, or simply referenced. Effort and attention will be concentrated on important issues; useless bulk will be eliminated.

- 40 CFR Part 50 concerning the National Ambient Air Quality Standards

- 40 CFR Part 51, Subpart W, concerning requirements related to applicable implementation plans

- 40 CFR Part 51, Appendix W, concerning air quality models

- 40 CFR Part 81, Subparts C and D, concerning attainment status designations approved by the EPA and identification of mandatory Class I Federal areas

- 40 CFR Part 93, Subpart B, concerning requirements for determining conformity of Federal actions to State or Federal implementation plans

Additional regulatory positions and specific criteria in support of the regulations identified above are as follows:

- Regulatory Guide 1.23, *Meteorological Monitoring Programs for Nuclear Power Plants* (NRC 2007)

- ESRP 2.7 in NUREG-1555 provides guidance on onsite meteorological measurements for use in licensing applications

Technical Rationale

The review conducted under this ESRP leads to preparation of a section describing the affected environment for the SEIS that provides background information to be used in evaluating air quality, meteorology, and noise impacts of continued plant operations and refurbishment activities associated with license renewal.

III. REVIEW PROCEDURES

The review of meteorology, air quality, and noise should be conducted in stages to permit termination when sufficient analysis has been completed to reach the appropriate conclusions. The following review steps are suggested:

1. Review the air quality discussion in the *Generic Environmental Impact Statement for License Renewal of Nuclear Power Plant* (GEIS; NUREG-1437) to identify the information considered and the conclusions reached. This step establishes the base for evaluation of information identified by the applicant, the public, and the staff.

2. Obtain descriptions of the site meteorological, climatological, dispersion characteristics, and acoustic (noise) environment from prior environmental documents.

3. Obtain recent meteorological data for the site and climatological data for the region surrounding the site.

4. Obtain the air-quality attainment status and available air-quality data for the region.

5. Update the descriptions of meteorology, climatology, air quality, and noise environment presented in prior environmental documents using recent data.

6. Determine if there is significant new information that would require specialized atmospheric modeling.

7. If the atmospheric dispersion calculations would be required to determine the potential impact of workers' vehicles on air quality in nonattainment or maintenance areas, or for dose calculations, or if specialized atmospheric modeling would be required, then continue the review at Step 9. Otherwise, prepare a section for the SEIS that presents an updated summary of the meteorology and climatology, air quality, and noise environment for the plant site and region. The summary should address normal conditions and historic severe weather, including the frequency and severity of severe weather phenomena.

8. Determine the specific atmospheric models to be used to support reviews being conducted under other ESRPs and the data requirements of those models. Models to support dose calculations are described in Regulatory Guides 1.111, Rev. 1, *Methods for Estimating Atmospheric Transport and Dispersion of Gaseous Effluents in Routine Releases from Light-Water-Cooled Reactors* (NRC 1977), and 1.145, Rev. 1, *Atmospheric Dispersion Models for Potential Accident Consequence Assessments at Nuclear Power Plants* (1983). Models approved by the Environmental Protection Agency for air quality calculations are listed in Appendix W of 40 CFR 51.

9. Prepare the meteorological data for use by the atmospheric models and assist the reviewers of the other ESRPs in making the appropriate model calculations.

10. Prepare a section for the supplemental environmental impact statement (SEIS) that presents an updated summary of the meteorology, climatology, air quality, and noise environment for the plant site and region. The summary should address normal conditions and historic severe weather. The section should describe and summarize the meteorological data used in atmospheric model calculations. The atmospheric models used should be identified in the SEIS, but detailed model descriptions should be avoided.

IV. EVALUATION FINDINGS

The reviewer should ensure that the meteorology, air quality, and noise information is adequate as a basis for assessment of the effects of continued plant operations and refurbishment associated with license renewal. The reviewer should consult with appropriate Federal, State, regional, local, and affected American Indian Tribes to assess the accuracy of the meteorology, air quality, and noise information, if necessary.

V. IMPLEMENTATION

The method described in this ESRP would be used by the staff in evaluating conformance with the Commission's regulations, except in those cases in which the applicant for license renewal proposes an acceptable alternative for complying with specified portions of the regulations.

VI. BIBLIOGRAPHY

10 CFR Part 51. *Code of Federal Regulations*, Title 10, *Energy*, Part 51, "Environmental Protection Regulations for Domestic Licensing and Related Regulatory Functions."

40 CFR Part 50. *Code of Federal Regulations*, Title 40, *Protection of Environment*, Part 50, "National Primary and Secondary Ambient Air Quality Standards."

40 CFR Part 51. *Code of Federal Regulations*, Title 40, *Protection of Environment*, Part 51, "Requirements for Preparation, Adoption, and Submittal of Implementation Plans."

40 CFR Part 81. *Code of Federal Regulations*, Title 40, *Protection of Environment*, Part 81, "Designation of Areas for Air Quality Planning Purposes."

U.S. Nuclear Regulatory Commission (NRC). 1977. *Methods for Estimating Atmospheric Transport and Dispersion of Gaseous Effluents in Routine Releases from Light-Water-Cooled Reactors*. Regulatory Guide 1.111, Rev. 1.

U.S. Nuclear Regulatory Commission (NRC). 1983. *Atmospheric Dispersion Models for Potential Accident Consequence Assessments at Nuclear Power Plants*. Regulatory Guide 1.145, Rev 1.

U.S. Nuclear Regulatory Commission (NRC). 2007. *Meteorological Monitoring Programs for Nuclear Power Plants*. Regulatory Guide 1.23.

U.S. Nuclear Regulatory Commission (NRC). 2013. *Generic Environmental Impact Statement for License Renewal of Nuclear Plants*. NUREG-1437, Vols. 1, 2, and 3, Revision 1. Office of Nuclear Reactor Regulation, Washington, D.C.

ENVIRONMENTAL STANDARD REVIEW PLAN

OFFICE OF NUCLEAR REACTOR REGULATION

3.3 GEOLOGIC ENVIRONMENT

I. AREAS OF REVIEW

This environmental standard review plan (ESRP) provides guidance for the review of the geology and soils of the site and surrounding area. This review should provide background information for inclusion in the supplemental environmental impact statement (SEIS) and input to reviewers for supplemental ESRPs for license renewal dealing with evaluation of the impacts of continued plant operations and refurbishment associated with license renewal.

The scope includes (1) description of geologic setting, (2) overview of seismicity and seismic history, (3) description of onsite soils and their relationship to site geology, and (4) description of soil erosion potential at the site.

Data and Information Needs

The types of data and information needed would be affected by nuclear power plant site- and plant-specific factors. The following data or information may be needed, if appropriate:

- descriptions of geologic setting at the plant site, including occurring rock types, formation names, and thicknesses

- descriptions of seismic potential at the site and seismic history

- identity of largest known local and historic regional earthquake

---

USNRC ENVIRONMENTAL STANDARD REVIEW PLAN

- description of safe-shutdown earthquake for the plant

- description of onsite soils (e.g. overburden and unconsolidated material) and their relationship to site geology (whether the material was brought in from offsite or is naturally occurring)

- description of onsite erosion control and run-off best management practices

- description of erosion potential at the site

- identity of any important farmland soils (e.g., prime farmland) on or in the vicinity of the site

- description of any rare or unique geologic resources, including rock, mineral, or energy rights and assets at or adjoining the site, including resource extraction activities

II. ACCEPTANCE CRITERIA

Acceptance criteria for the evaluation of site geology and soils are based on the following regulations:

- 10 CFR 51.45(d), "Status of Compliance." The environmental report shall list all Federal permits, licenses, approvals and other entitlements which must be obtained in connection with the proposed action and shall describe the status of compliance with these requirements. The environmental report shall also include a discussion of the status of compliance with applicable environmental quality standards and requirements including, but not limited to, applicable zoning and land-use regulations, and thermal and other water pollution limitations or requirements which have been imposed by Federal, State, regional, and local agencies having responsibility for environmental protection.

- 10 CFR 51.53(c)(2). The report must contain a description of the proposed action, including the applicant's plans to modify the facility or its administrative control procedures as described in accordance with 10 CFR 54.21 of this chapter. This report must describe in detail the affected environment around the plant, the modifications directly affecting the environment or any plant effluents, and any planned refurbishment activities. In addition, the applicant shall discuss in this report the environmental impacts of alternatives and any other matters discussed in 10 CFR 51.45.

- 10 CFR 51.70(b). The draft environmental impact statement will be concise, clear, and analytic, and written in plain language with appropriate graphics....The format provided in Section 1(a) of Appendix A of this subpart should be used. The Nuclear Regulatory Commission (NRC) staff will independently evaluate and be responsible for the reliability of all information used in the draft environmental impact statement.

- 10 CFR Part 51, Appendix A to Subpart A, para. 6, concerning affected environment. The environmental impact statement will succinctly describe the environment to be affected by the proposed action. Data and analyses in the statement will be commensurate with the importance of the impact, with less important material summarized, consolidated, or simply referenced. Effort and attention will be concentrated on important issues; useless bulk will be eliminated.

<u>Technical Rationale</u>

The review conducted under this ESRP leads to preparation of a section describing the affected environment for the SEIS that provides background information to be used in evaluating impacts associated with geology and soils from continued plant operations and refurbishment activities associated with license renewal.

III. REVIEW PROCEDURES

The review of geology and soils should be conducted in stages to permit termination when sufficient analysis has been completed to reach the appropriate conclusions. The following review steps are suggested:

1. Review the discussion of potential impacts of continued plant operation and refurbishment activities on geology and soils in the *Generic Environmental Impact Statement for License Renewal of Nuclear Power Plant* (GEIS; NUREG-1437) to identify the information considered and the conclusions reached. This step establishes the base for evaluation of information identified by the applicant, the public, and the staff.

2. Obtain descriptions of regional and local geology, soils, geologic resources, and seismic setting.

3. Obtain descriptions of the site geology, soils, geologic resources, and seismic setting from prior environmental documents.

4. Obtain descriptions of seismic potential at the site and seismic history, including the largest known local and historic regional earthquake and safe-shutdown earthquake for the plant.

5. Obtain descriptions of any onsite erosion control plans and run-off best management practices.

6. Prepare a section for the SEIS that presents an updated summary of the geology and soils, including significant geologic resources, and seismic setting for the plant site and surrounding region.

IV. EVALUATION FINDINGS

The reviewer should ensure that the geology and soils information is adequate as a basis for assessment of the effects of continued plant operations and refurbishment associated with license renewal. The reviewer should consult with appropriate Federal, State, regional, and local agencies, as well as Indian Tribes, to assess the accuracy of the geology and soils information, if necessary.

V. IMPLEMENTATION

The method described in this ESRP would be used by the staff in evaluating conformance with the Commission's regulations, except in those cases in which the applicant for license renewal proposes an acceptable alternative for complying with specified portions of the regulations.

VI. BIBLIOGRAPHY

10 CFR Part 51. *Code of Federal Regulations*, Title 10, *Energy,* Part 51, "Environmental Protection Regulations for Domestic Licensing and Related Regulatory Functions."

40 CFR Part 51. *Code of Federal Regulations*, Title 40, *Protection of Environment*, Part 51, "Requirements for Preparation, Adoption, and Submittal of Implementation Plans."

U.S. Nuclear Regulatory Commission (NRC). 2013. *Generic Environmental Impact Statement for License Renewal of Nuclear Plants.* NUREG-1437, Vols. 1, 2, and 3, Revision 1. Office of Nuclear Reactor Regulation, Washington, D.C.

U.S. NUCLEAR REGULATORY COMMISSION

ENVIRONMENTAL STANDARD
REVIEW PLAN

OFFICE OF NUCLEAR REACTOR REGULATION

3.4 WATER RESOURCES

I. AREAS OF REVIEW

This environmental standard review plan (ESRP) provides guidance for the review of water use and quality that could affect or be affected by continued plant operations and refurbishment associated with license renewal.

The scope includes (1) consideration of such water uses as domestic, municipal, agricultural, industrial, mining, recreation, navigation, and hydroelectric power, (2) identification of their locations, (3) quantification of water diversions, consumption, and returns, (4) consideration of site-specific and regional data on the physical, chemical, and biological characteristics of groundwater and surface water to provide the basic data for the evaluation of water-quality impacts to water bodies, aquifers, aquatic ecosystems, (5) water use related to continued plant operations and refurbishment associated with license renewal, and (6) preparation of a section describing water use and water quality for the supplemental environmental impact statement (SEIS). The review should be limited to present and reasonably known future water uses.

Data and Information Needs

The types of data and information needed would be affected by nuclear power plant site- and plant-specific factors. The following data or information may be needed:

USNRC ENVIRONMENTAL STANDARD REVIEW PLAN

- maps (including digital databases such as a Geographic Information System) showing the relationship of the site to the major hydrological systems, surface water bodies, floodplains, and groundwater aquifer systems that could affect or be affected by plant water withdrawals and/or discharges

- quantitative descriptions of present and known future surface water uses (withdrawals, consumptions, and returns), groundwater withdrawals, and nonconsumptive water uses (recreational, navigational, instream, etc.) that may affect or be affected by continued plant operations and refurbishment associated with license renewal. This should include any bodies of water or aquifers at distances close enough to affect or be adversely affected by the plants. This should also include a quantitative description of any water uses that provide potential liquid pathways for both radiological and nonradiological effluents. The following should be included:

    - locations of diversions and returns concerning the site and the water body

    - identification of the water body

    - average monthly withdrawal and return rate for each surface water diversion by use category.

    - locations and depths of wells in relation to the site

    - identification of aquifers

    - identification of any Environmental Protection Agency-designated sole source aquifers

    - the average monthly groundwater withdrawal rates by use category

    - identification of water bodies and locations concerning the site within a 6-mile (10-kilometer) radius, including any delineated floodplains or zones of inundation for adjoining and onsite surface water features (maps may be useful)

    - the type and location of activity on the identified water body (maps may be useful)

    - the use rate variation over time

- summary of statutory and other legal restrictions relating to water use or specific water-body restrictions on water use imposed by Federal or State regulations

- a water-use diagram for the plant showing flow rates to and from the various water systems (e.g., circulating water system, sanitary system, radwaste and chemical waste systems, service water systems), points of consumption, and source and discharge locations

- for the water-use diagram, the data and narrative description for maximum water consumption, water consumption during periods of minimum water availability, and average operation by month and by plant operating status

- a description of other station water uses (i.e., all facilities not associated with the proposed plant) showing flow rates to and from the facility, average water consumption, and maximum water consumption

- the mean, range, and temporal and spatial variations of the surface water and groundwater quality characteristics

  - For surface waters: water temperature, suspended solids, total dissolved solids, hardness, turbidity, color, odor, conductivity, dissolved oxygen, biological oxygen demand, chemical oxygen demand, phosphorus forms (total and orthophosphate), nitrogen forms (ammonia, nitrate, nitrite, organic), alkalinity, chlorides, sulfate, sodium, potassium, calcium, magnesium, heavy metals (e.g., Hg, Pb), pH, phytoplankton (chlorophyll a), and indicator microorganisms (e.g., total coliform, fecal coliforms, fecal streptococci)

  - For groundwaters: the above-surface-water data, minus phytoplankton and with silica, iron, and bicarbonate added

- other nuclear power plant site-specific water-quality characteristics

- descriptions of preexisting aquatic environmental stresses and their effects on surface or groundwater quality for waters that interact with the plant (e.g., water bodies at or near the site that do not meet established water-quality standards). These should include State 303d lists of impaired waters which classify the quality of each of the State's water bodies and are required under the Federal Water Pollution Control Act (Clean Water Act)

- descriptions of pollutant sources with discharges to water that may interact with the plant, including locations relative to the site and the affected water bodies, and the magnitude and nature of the pollutant discharges, including spatial and temporal variations

II. ACCEPTANCE CRITERIA

Acceptance criteria for the review of water use are based on the following regulations:

- 10 CFR 51.45(d), "Status of Compliance." The environmental report shall list all Federal permits, licenses, approvals and other entitlements which must be obtained in connection with the proposed action and shall describe the status of compliance with these requirements. The environmental report shall also include a discussion of the status of compliance with applicable environmental quality standards and requirements including, but not limited to, applicable zoning and land-use regulations, and thermal and other water pollution limitations or requirements which have been imposed by Federal, State, regional, and local agencies having responsibility for environmental protection.

- 10 CFR 51.53(c)(2). The report must contain a description of the proposed action, including the applicant's plans to modify the facility or its administrative control procedures as described in accordance with 10 CFR 54.21 of this chapter. This report must describe in detail the affected environment around the plant, the modifications directly affecting the environment or any plant effluents, and any planned refurbishment activities. In addition, the applicant shall discuss in this report the environmental impacts of alternatives and any other matters discussed in 10 CFR 51.45.

- 10 CFR 51.53(c)(3)(ii)(A). If the applicant's plant utilizes cooling towers or cooling ponds and withdraws makeup water from a river, an assessment of the impact of the proposed action on water availability and competing water demands, the flow of the river, and related impacts on stream (aquatic) and riparian (terrestrial) ecological communities must be provided. The applicant shall also provide an assessment of the impacts of the withdrawal of water from the river on alluvial aquifers during low flow.

- 10 CFR 51.53(c)(3)(ii)(C). If the applicant's plant pumps more than 100 gallons (total onsite) of groundwater per minute, an assessment of the impact of the proposed action on groundwater must be provided.

- 10 CFR 51.53(c)(3)(ii)(D). If the applicant's plant is located at an inland site and utilizes cooling ponds, an assessment of the impact of the proposed action on groundwater quality must be provided.

- 10 CFR 51.53(c)(3)(ii)(P). An applicant shall assess the impact of any documented inadvertent releases of radionuclides into groundwater. The applicant shall include in its assessment a description of any groundwater protection program used for the surveillance of piping and components containing radioactive liquids for which a pathway to groundwater may exist. The assessment must also include a description of any past inadvertent releases and the projected impact to the environment (e.g., aquifers, rivers, lakes, ponds, the ocean) during the license renewal term.

- 10 CFR 51.70(b). The draft environmental impact statement will be concise, clear, and analytic, and written in plain language with appropriate graphics....The format provided in Section 1(a) of Appendix A of this subpart should be used. The NRC staff will independently evaluate and be responsible for the reliability of all information used in the draft environmental impact statement.

- 10 CFR Part 51, Appendix A to Subpart A, para. 6, concerning affected environment. The environmental impact statement will succinctly describe the environment to be affected by the proposed action. Data and analyses in the statement will be commensurate with the importance of the impact, with less important material summarized, consolidated, or simply referenced. Effort and attention will be concentrated on important issues; useless bulk will be eliminated.

- 33 CFR Part 330, Appendix A, concerning conditions, limitations, and restrictions on construction activities

- 40 CFR Parts 122-133, concerning National Pollutant Discharge Elimination System (NPDES) permit conditions for discharges, including storm-water discharges

- 40 CFR Part 147, concerning restrictions on waste disposal options

- 40 CFR Part 149, concerning possible supplemental restrictions on waste disposal and water use in or above a sole source aquifer

- 40 CFR Part 165, concerning the disposal and storage of pesticides and pesticide containers

- 40 CFR Part 403, concerning waste effluents

- 40 CFR Part 423, concerning effluent limitations on existing and new point sources

- 40 CFR Parts 700–716, concerning practices and procedures for managing toxic chemicals

- Federal, State, regional, local, and Indian Tribal water laws and water rights

Additional regulatory positions and specific criteria in support of regulations identified above are as follows:

- Compliance with environmental quality standards and requirements of the Clean Water Act is not a substitute for and does not negate the requirement for the Nuclear Regulatory Commission (NRC) to weigh the environmental impacts of the proposed action, including any degradation of water quality, and to consider alternatives to the proposed action that are available for reducing the adverse impacts. If an environmental assessment of aquatic impacts is available from the permitting authority, the NRC would consider the assessment in its determination of the magnitude of the environmental impacts in striking an overall cost-benefit balance. When no such assessment of aquatic impacts is available from the permitting authority, the NRC (possibly in conjunction with the permitting authority and other agencies having relevant expertise) should establish its own impact determination.

- Because water quality and water supply are interdependent, changes in water quality must be considered simultaneously with changes in water supply. In *PUD No. 1 of Jefferson County v. Washington Department of Ecology*, 511 U.S. 700 (1994), the United States Supreme Court interpreted the Clean Water Act as allowing States to impose conditions on certifications, such as limitations on a given project, insofar as necessary to enforce a designated use contained in the State's water quality standard. The Court held that these limitations do not have to be specifically tied to a discharge requirement.

Technical Rationale

The review conducted under this ESRP leads to preparation of a section describing the affected environment for the SEIS that provides background information to be used in evaluating water use and quality impacts of continued plant operations and refurbishment activities associated with license renewal.

III. REVIEW PROCEDURES

The reviewer's analysis of surface water and groundwater use should consider the aspects of water use that are concerned with consumptive use, nonconsumptive use, and effluent pathways. The depth of analysis would be related to the importance of water use and proximity of the use to the plant. Emphasis should be given to those water uses that are expected to change during refurbishment and/or the license renewal term. The following review steps are suggested:

1. Identify consumptive water uses that could affect the water supply of the plant or that may be adversely affected by the plant, including the following important characteristics:

    *   water source

    *   locations of diversions and returns

    *   amount and time variation of use

    *   water rights

2. Identify recreational, navigational, and other nonconsumptive water uses. The important characteristics to be specified are:

    *   location

    *   activity

    *   amount and time variation of use

3. Identify the water uses that provide potential pathways for both radiological and nonradiological effluents, including the following important characteristics:

    *   water sources

    *   location of diversions for consumptive uses

    *   location of receptors for nonconsumptive uses

    *   amount and time variation of use for each

4. In addition to information obtained from the applicant's environmental report (ER) and from responses to subsequent questions to the applicant, use additional sources of data, such as

    *   local water-supply companies or agencies

    *   river basin commissions

    *   State agencies (e.g., water resources, fish and wildlife)

    *   various Federal agencies, such as the U.S. Army Corps of Engineers and the U.S. Geological Survey, and Indian Tribal agencies when needed to complete the analysis. Local water users may be questioned during the site visit.

5. Using the above information, compile and tabulate water uses by the categories and characteristics described in this ESRP section, but limit the analysis to consideration of present and reasonably known future water uses.

6. Ensure that water-use data and information are adequate to serve as a basis for assessing the impacts of continued plant operations and refurbishment associated with license renewal on water use.

- In evaluating the adequacy of this material, the reviewer should ensure that data are sufficient to predict water-use impacts to the plant as well as water-use characteristics to be impacted by refurbishment and operation during the renewal term.

- Consult with appropriate Federal, State, regional, local, and Tribal agencies in making this evaluation.

The reviewer's analysis of water quality should ensure that the physical, chemical, and biological water-quality parameters that could affect or be affected by continued plant operations during the license renewal term and refurbishment in support of license renewal have been described. The reviewer should take the following steps:

1. Identify the location and spatial distribution of the physical, chemical, and biological characteristics, the monthly and annual ranges, and the historical extremes of those water-quality characteristics that could potentially affect or be affected by continued plant operations and refurbishment.

2. Determine the presence of existing water-quality-related environmental stresses. Consult the quality criteria requirements of other water users, as indicated by the approved water-use classification (such as 303(d), lists) or water resource planning documents for the water body in question.

3. When applicable, discuss the water-quality conditions, floodplains and waterway buffer zones, water rights, and agreements as they affect water quality and water supply and resource plans for the site and vicinity with Federal, State, regional, local, and affected Native American Tribal water resource and pollution control and monitoring agencies.

4. Obtain the information primarily from the applicant's ER, responses to questions to the applicant, and consultation with Federal, State, regional, local, and affected Native American Tribal agencies. Use sources of data such as river basin planning organizations and State and Federal agencies, such as the Environmental Protection Agency, the U.S. Army Corps of Engineers, and the U.S. Geological Survey, if additional information or verification is deemed necessary.

5. Ensure that the

   - data are sufficient to provide quantitative information on the physical, chemical, and biological water-quality characteristics potentially affecting or affected by continued plant operations and refurbishment

   - hydrologic and water-quality descriptions are sufficient, concerning relevancy, completeness, reliability, and accuracy for input to the impact assessments of other sections

   - Federal, State, regional, local, and Tribal agencies appropriate to the objectives of this review have been consulted.

6. When evaluating the adequacy of this material:

   - consult the applicable standards and guides for this environmental review and use the site visit and/or consultations with permitting agencies to evaluate the completeness of the water-quality descriptions.

   - evaluate, when necessary, the collection of additional data, the verification of data, and the substantiation of the methodology used to estimate water-quality parameters

7. Include the appropriate depth and extent of the input to the environmental impact statement (EIS) as governed by the hydrologic and water-quality characteristics that could affect or be affected by continued plant operations and refurbishment and by the nature and magnitude of the expected impacts. The following information should be included as input to the EIS:

   - descriptions of site and vicinity surface-water and groundwater occurrence, flow, and quality that could affect or be affected by continued plant operations and refurbishment. The description may consist of statistical summaries of the relevant characteristics, including mean, mean low and high, and historical low and high values (as available) for the site and vicinity. The data included should be commensurate with the anticipated impacts. Figures may be used to show long-term and seasonal trends, such as variations in dissolved oxygen and nutrient concentrations and pH variations.

   - a description of the water-quality related environmental stresses in the site and vicinity.

IV. EVALUATION FINDINGS

The reviewer should ensure that the water-use information is adequate as a basis for assessment of the effects of continued plant operations and refurbishment associated with license renewal. The depth and extent of the input to the SEIS would be governed by the water-use characteristics of the site and vicinity and the potential water-use impacts of continued plant operations during the license renewal term and refurbishment in support of license renewal. The information should be presented in a concise form. Data should be given in tables where appropriate.

The following information should be included:

- a summary of present and reasonably known future groundwater withdrawals on the site and for distances great enough to cover potentially affected groundwater aquifers

- a summary of present and reasonably known future surface-water uses that are within the hydrological system in which the plant is located and that may affect or be adversely affected by the plant

- a summary of present and reasonably known future recreational, navigational, and other nonconsumptive water uses (maps may be useful)

- references to applicable Federal, State, regional, local, and Tribal water use laws

V. IMPLEMENTATION

The method described in this ESRP would be used by the staff in evaluating conformance with the Commission's regulations, except in those cases in which the applicant for license renewal proposes an acceptable alternative for complying with specified portions of the regulations.

VI. BIBLIOGRAPHY

10 CFR Part 51. *Code of Federal Regulations*, Title 10, *Energy,* Part 51, "Environmental Protection Regulations for Domestic Licensing and Related Regulatory Functions."

33 CFR Part 330. *Code of Federal Regulations*, Title 33, *Navigation and Navigable Waters*, Part 330, "Nationwide Permit Program."

40 CFR Part 122. *Code of Federal Regulations*, Title 40, *Protection of Environment*, Part 122, "EPA Administered Permit Programs: The National Pollutant Discharge Elimination System."

40 CFR Part 123. *Code of Federal Regulations*, Title 40, *Protection of Environment*, Part 123, "State Program Requirements."

40 CFR Part 124. *Code of Federal Regulations*, Title 40, *Protection of Environment*, Part 124, "Procedures for Decisionmaking."

40 CFR Part 125. *Code of Federal Regulations*, Title 40, *Protection of Environment*, Part 125, "Criteria and Standards for the National Pollutant Discharge Elimination System."

40 CFR Part 129. *Code of Federal Regulations*, Title 40, *Protection of Environment*, Part 129, "Toxic Pollutant Effluent Standards."

40 CFR Part 130. *Code of Federal Regulations*, Title 40, *Protection of Environment*, Part 130, "Water Quality Planning and Management."

40 CFR Part 131. *Code of Federal Regulations*, Title 40, *Protection of Environment*, Part 131, "Water Quality Standards."

40 CFR Part 132. *Code of Federal Regulations*, Title 40, *Protection of Environment*, Part 132, "Water Quality Guidance for the Great Lakes System."

40 CFR Part 133. *Code of Federal Regulations*, Title 40, *Protection of Environment*, Part 133, "Secondary treatment Regulation."

40 CFR Part 147. *Code of Federal Regulations*, Title 40, *Protection of Environment*, Part 147, "State, Tribal, and EPA-Administered Underground Injection Control Programs."

40 CFR Part 149. *Code of Federal Regulations*, Title 40, *Protection of Environment*, Part 149, "Sole Source Aquifers."

40 CFR Part 165. *Code of Federal Regulations*, Title 40, *Protection of Environment*, Part 165, "Pesticide Management and Disposal."

40 CFR Part 403. *Code of Federal Regulations*, Title 40, *Protection of Environment*, Part 403, "General Pretreatment Regulations for Existing and New Sources of Pollution."

40 CFR Part 423. *Code of Federal Regulations*, Title 40, *Protection of Environment*, Part 423, "Steam Electric Power Generating Point Source Category."

40 CFR Part 700. *Code of Federal Regulations*, Title 40, *Protection of Environment*, Part 700, "General."

40 CFR Part 702. *Code of Federal Regulations*, Title 40, *Protection of Environment*, Part 702, "General Practices and Procedures."

40 CFR Part 704. *Code of Federal Regulations*, Title 40, *Protection of Environment*, Part 704, "Reporting and Recordkeeping Requirements."

40 CFR Part 707. *Code of Federal Regulations*, Title 40, *Protection of Environment*, Part 707, "Chemical Imports and Exports."

40 CFR Part 710. *Code of Federal Regulations*, Title 40, *Protection of Environment*, Part 710, "Compilation of the TSCA Chemical Substance Inventory."

40 CFR Part 711. *Code of Federal Regulations*, Title 40, *Protection of Environment*, Part 711, "TSCA Chemical Data Reporting Requirements."

40 CFR Part 712. *Code of Federal Regulations*, Title 40, *Protection of Environment*, Part 712, "Chemical Information Rules."

40 CFR Part 716. *Code of Federal Regulations*, Title 40, *Protection of Environment*, Part 716, "Health and Safety Data Reporting."

Federal Water Pollution Control Act (FWPCA), as amended, 33 USC 1251 et seq. (Clean Water Act).

PUD No. 1 of Jefferson County v. Washington Department of Ecology, 511 U.S. 700 (1994)

Rosaler, R. (ed). 1994. *Standard Handbook of Plant Engineering*. Second Edition, McGraw-Hill, New York.

U.S. Nuclear Regulatory Commission (NRC). 2013. *Generic Environmental Impact Statement for License Renewal of Nuclear Plants.* NUREG-1437, Vols. 1, 2, and 3, Revision 1. Office of Nuclear Reactor Regulation, Washington, D.C.

U.S. NUCLEAR REGULATORY COMMISSION
ENVIRONMENTAL STANDARD REVIEW PLAN
OFFICE OF NUCLEAR REACTOR REGULATION

3.5 ECOLOGICAL RESOURCES

I. AREAS OF REVIEW

This environmental standard review plan (ESRP) provides guidance for the review of terrestrial and aquatic resources that could affect or be affected by continued plant operations and refurbishment associated with license renewal.

The scope of the terrestrial review should include (1) a description of species composition, spatial and temporal distribution, abundance, and other structural and functional attributes of biotic assemblages that could be affected by license renewal; (2) identification and location of any vulnerable, irreplaceable, or otherwise important terrestrial natural resources, habitats, and wildlife sanctuaries and preserves; and (3) preparation of a section describing terrestrial resources for the supplemental environmental impact statement (SEIS).

The scope of the aquatic review should include (1) a description of the spatial and temporal distribution, abundance, and other structural and functional attributes of biotic assemblages that could be affected by license renewal, (2) identification of any "important" or irreplaceable aquatic natural resources and the location of sanctuaries and preserves, and (3) preparation of a section describing aquatic resources for the SEIS.

Data and Information Needs

Information and data on terrestrial and aquatic ecosystems and resources should be utilized for developing the following descriptions of ecological resources in the SEIS.

- Region, describe:

3.5-1

USNRC ENVIRONMENTAL STANDARD REVIEW PLAN

Environmental standard review plans are prepared for the guidance of the Office of Nuclear Reactor Regulation staff responsible for environmental reviews for nuclear power plants. These documents are made available to the public as part of the Commission's policy to inform the nuclear industry and the general public of regulatory procedures and policies. Environmental standard review plans are not substitutes for regulatory guides or the Commission's regulations and compliance with them is not required. These supplemental environmental standard review plans are keyed to Regulatory Guide 4.2, Supplement 1, "Preparation of Environmental Reports for Nuclear Power Plant License Renewal Applications."

Published environmental standard review plans will be revised periodically, as appropriate, to accommodate comments and to reflect new information and experience.

Comments and suggestions for improvement will be considered and should be sent to the U.S. Nuclear Regulatory Commission, Office of Nuclear Reactor Regulation, Washington, DC 20555-0001.

- the ecosystem and habitats of the ecoregion that includes the site
- the geomorphic, or physiographic, province
- the characteristic vegetation and animal species, including climax vegetation and typical succession in the area of the site
- the marine ecoregion of the ocean or gulf waters near the plant if it is located near such waters
- the watershed(s) and names and locations of source and receiving water bodies for the plant's cooling system. Provide the U.S. Department of Agriculture (USDA) Plant Hardiness Zone.

- Site and vicinity, describe:

    - local soil types
    - water and sediment quality
    - vegetation and animal communities
    - physiographic habitats (such as upland forest, swamps marshes, wetlands, rivers, streams, etc.)
    - significant water bodies that intercept or parallel in-scope transmission lines
    - Include detailed maps and descriptions, as appropriate that contain site and in-scope transmission lines and stream crossings, any significant natural heritage areas, and known locations or historic sightings of migratory bird rookeries, etc.

- Potentially affected water bodies, describe: (Aquatic Ecology)

    - location of site concerning the principal nearby water bodies that it affects
    - the source and receiving water bodies in terms of relationship to the watershed, size, shoreline, bathymetry, tidal and net flows including seasonal or occasional variations, bottom types (e.g., sand, silt, clay, sandy silt, silty sand, etc.), and both sediment and water quality
    - location of main channel, dams, and flood control
    - additional uses of the water body other than for cooling water.

- In-scope transmission lines and rights-of-way (ROWs), describe: (Terrestrial Ecology)

    - transmission lines and ROWs including length of transmission lines or portions of lines and width of ROWs
    - ROW maintenance plans, procedures, or protocols and pesticides and herbicides used in ROW maintenance, including descriptions of how and when they are released
    - locations where ROWs cross wetlands and waterfowl areas as well as other vulnerable or otherwise important terrestrial habitats (also map these areas)

Additionally, the reviewer should provide brief descriptions of the following:

- the environment before European settlement and the transition of the environment on the site from before plant construction to the present. Include major changes or modifications to the land and/or water bodies over the projected life of the plant. Typical things to describe include channelization, navigation, pollution, habitat degradation or fragmentation, urbanization, development, and pond or reservoir creation. Include pollution control or other programs designed for environmental improvement.

- major wildlife living around site in the past and which species remain today. Refer to historic and cultural resources when possible to avoid repeating information.

- occurrence, location, and description of communities and habitats of special interest in the vicinity of plant, such as wetlands, natural heritage areas and other areas of public or scientific interest, or others that may be particularly sensitive or susceptible either directly or indirectly to the effects of continued plant operations during the license renewal term and refurbishment in support of license renewal

- occurrence of invasive species in the vicinity of the plant and documentation of any management activities undertaken by the plant to control such species

- wildlife management plans and best management practices (if applicable), including pesticides and herbicides used routinely to maintain the site

The reviewer should summarize the following, if applicable:

- studies or monitoring programs of ecological resources on or in the vicinity of the site in terms of when and where conducted and by whom, the objective (why conducted), the biological entities or attributes chosen for study, methods and results (how conducted) applicable to this license renewal application, and what data or data summaries might be available for Nuclear Regulatory Commission (NRC) review

- information on Federally-listed threatened, endangered, or candidate species and critical habitat and State-listed species and habitat that could be found on the site or in the vicinity of the site, including the area within the in-scope transmission line corridors. For such species, summarize historical occurrences, population size and trends, critical habitat, and potential habitat. Similar information will be needed for any species with essential fish habitat on or in the vicinity of the site. Describe any license renewal and refurbishment activities as well as any modifications to plant operation that may affect such species and habitats.

II. ACCEPTANCE CRITERIA

Acceptance criteria for the review of ecological resources are based on the following regulations:

- 10 CFR 51.45(d), "Status of Compliance." The environmental report shall list all Federal permits, licenses, approvals and other entitlements which must be obtained in connection with the proposed action and shall describe the status of compliance with these requirements. The environmental report shall also include a discussion of the status of compliance with applicable environmental quality standards and requirements including, but not limited to, applicable zoning and land-use regulations, and thermal and other water pollution limitations or requirements which have been imposed by Federal, State, regional, and local agencies having responsibility for environmental protection.

- 10 CFR 51.53(c)(2). The report must contain a description of the proposed action, including the applicant's plans to modify the facility or its administrative control procedures as described in accordance with 10 CFR 54.21 of this chapter. This report must describe in detail the affected environment around the plant, the modifications directly affecting the environment or any plant effluents, and any planned refurbishment activities. In addition, the applicant shall discuss in this report the environmental impacts of alternatives and any other matters discussed in 10 CFR 51.45.

- 10 CFR 51.53(c)(3)(ii)(A). If the applicant's plant utilizes cooling towers or cooling ponds and withdraws makeup water from a river, an assessment of the impact of the proposed action on water availability and competing water demands, the flow of the river, and related impacts on stream (aquatic) and riparian (terrestrial) ecological communities must be provided.

- 10 CFR 51.53(c)(3)(ii)(B). If the applicant's plant utilizes once-through cooling or cooling pond heat dissipation systems, the applicant shall provide a copy of current Clean Water Act 316(b) determinations and, if necessary, a 316(a) variance in accordance with 40 CFR Part 125 or equivalent State permits and supporting documentation. If the applicant cannot provide these documents, it shall assess the impact of the proposed action on fish and shellfish resources resulting from thermal changes and impingement and entrainment.

- 10 CFR 51.53(c)(3)(ii)(E). All license renewal applicants shall assess the impact of refurbishment, continued operations, and other license-renewal-related construction activities on important plant and animal habitats. Additionally, the applicant shall assess the impact of the proposed action on threatened or endangered species in accordance with Federal laws protecting wildlife, including but not limited to the Endangered Species Act, and essential fish habitat in accordance with the Magnuson-Stevens Fishery Conservation and Management Act.

- 10 CFR 51.70(b). The draft environmental impact statement will be concise, clear, and analytic, and written in plain language with appropriate graphics....The format provided in Section 1(a) of Appendix A of this subpart should be used. The NRC staff will independently evaluate and be responsible for the reliability of all information used in the draft environmental impact statement.

- 10 CFR Part 51, Appendix A to Subpart A, para. 6, concerning affected environment. The environmental impact statement will succinctly describe the environment to be affected by the proposed action. Data and analyses in the statement will be commensurate with the importance of the impact, with less important material summarized, consolidated, or simply referenced. Effort and attention will be concentrated on important issues; useless bulk will be eliminated.

- Endangered Species Act of 1973 (16 USC 1531, et seq.), requires the identification of threatened and endangered species and critical habitats, and formal or informal consultation with the U.S. Fish and Wildlife Service and/or National Marine Fisheries Service.

- Bald and Golden Eagle Protection Act (16 USC 668 et seq.), prohibits the taking, possessing, selling, purchasing, bartering, transporting, importing, or exporting of the bald or golden eagle, dead or alive, or any part, nest, or egg without a permit.

- Migratory Bird Treaty Act (16 USC 703, et seq.), concerning declaring that it is unlawful to take, import, export, possess, buy, sell, purchase, or barter any migratory bird. Feathers, or other parts of nests and eggs, and products made from migratory birds are also covered by the Act. "Take" is defined as pursuing, hunting, shooting, poisoning, wounding, killing, capturing, trapping, or collecting.

- Coastal Zone Management Act of 1972 (16 USC 1451 et seq.), requires that Federal actions, including licensing or permitting actions, be consistent with U.S. Department of Commerce (NOAA) approved State Coastal Management Plans, which protect resources in a state's coastal zone.

- Clean Water Act (33 USC 1251, et seq.), concerning restoration and maintenance of the chemical, physical, and biological integrity of water resources.

- Marine Mammal Protection Act of 1972 (16 USC 1361, et seq.), concerning the protection of marine mammals.

- Marine Protection, Research, and Sanctuaries Act of 1972 (33 USC 1401, et seq.), concerning dumping of dredged material into the ocean.

- Rivers and Harbors Appropriations Act of 1899 (33 USC 403), concerning the deposition of debris in navigable waters, or tributaries to such waters.

- The Magnuson-Stevens Fishery Conservation and Management Act, as amended (16 USC 1801, et seq.), established the Essential Fish Habitat provisions to identify and protect important habitats of federally managed marine and anadromous fish species and shellfish species.

Additional regulatory positions and specific criteria in support of regulations identified above are as follows:

- Regulatory Guide 4.11, Rev. 2, *Terrestrial Environmental Studies for Nuclear Power Stations* (NRC 2012), contains technical information for the design and execution of terrestrial environmental studies, the results of which may be appropriate for inclusion in the applicant's environmental report. The reviewer should ensure that the appropriate results are included in the environmental report.

Technical Rationale

The review conducted under this ESRP leads to preparation of a section describing the affected environment for the SEIS that provides background information to be used in evaluating terrestrial and aquatic resource impacts of continued plant operations and refurbishment activities associated with license renewal.

This description of the environment includes the ecological resources potentially affected by current and future nuclear plant operations. Ecological resources include members and attributes of aquatic, terrestrial, riparian, and wetland plant and animal communities. Wetlands and riparian habitats are the interface between aquatic and terrestrial habitats and are defined as follows. NRC generally includes wetland and riparian habitats under terrestrial resources. According to the Environmental Protection Agency (1993):

> Wetlands are defined as–
>
> "Those areas that are inundated or saturated by surface or ground water at a frequency and duration sufficient to support, and that under normal circumstances do support, a prevalence of vegetation typically adapted for life in saturated soil conditions. Wetlands generally include swamps, marshes, bogs, and similar areas."
>
> Riparian areas are defined as–
>
> "Vegetated ecosystems along a water body through which energy, materials, and water pass. Riparian areas characteristically have a high water table and are subject to periodic flooding and influence from the adjacent water body. These systems encompass wetlands, uplands, or some combination of these two land forms; they do not in all cases have all of the characteristics necessary for them to be classified as wetlands."

III. REVIEW PROCEDURES

The reviewer should ensure that the regional and site-specific terrestrial and aquatic ecological information is adequate to serve as a basis for assessment of the effects of continued plant operations and refurbishment associated with license renewal. The following review steps are suggested:

Terrestrial Review:

1. Describe the terrestrial communities from available information such as present and past studies, Federal and state sources, etc., and include representative species of plants, mammals, birds, reptiles, and amphibians.

2. Note any endemic species, sensitive or indicator species, or keystone species.

3. Include a description of bird species that nest within the area, migratory species, known migratory bird rookeries, and, if applicable, location of the site in relation to any nearby flyways.

4. Describe the types of vegetative communities found on and in the vicinity of the site, especially any delineated wetlands or potential wetland habitat.

5. Include a summary of any available botanical and wildlife surveys that have been conducted on or in the vicinity of the site.

Aquatic Review:

1. Describe the aquatic communities from available information such as present and past studies, Federal and State sources, etc.

2. Focus on a subset of representative and important species, such as those with the following characteristics: potential or reported susceptibility to impingement and entrainment; dominance, commonness, or rarity in numbers or biomass; importance to the structure and function of the ecosystem, such as keystone species, important trophic links, potential for trophic cascade, or habitat formers or modifiers; indicators of water quality or "ecosystem health"; important recreational or commercial fish and shellfish; fish consumption advisories; and ecosystem services.

IV. EVALUATION FINDINGS

The reviewer should ensure that the ecological information is adequate as a basis for assessment of the effects of continued plant operations and refurbishment associated with license renewal. The reviewer should consult with appropriate Federal and State agencies to assess the accuracy of the ecological information, if necessary.

V. IMPLEMENTATION

The method described in this ESRP would be used by the staff in evaluating conformance with the Commission's regulations, except in those cases in which the applicant for license renewal proposes an acceptable alternative for complying with specified portions of the regulations.

VI. BIBLIOGRAPHY

10 CFR Part 51. *Code of Federal Regulations*, Title 10, *Energy*, Part 51, "Environmental Protection Regulations for Domestic Licensing and Related Regulatory Functions."

Bald and Golden Eagle Protection Act of 1940, as amended. 16 USC 668 et seq.

Coastal Zone Management Act of 1972. 16 USC 1451 et seq.

Endangered Species Act. 16 USC 1531 et seq.

Executive Order 11988, "Floodplain Management."

Executive Order 11990, "Protection of Wetlands."

Federal Water Pollution Control Act of 1977 (Clean Water Act). 33 USC 1251, et seq.

Magnuson-Stevens Fishery Conservation and Management Act. 16 USC 1801 et seq.

Marine Mammal Protection Act of 1972. 16 USC 1361, et seq.

Marine Protection, Research, and Sanctuaries Act of 1972. 33 USC 1401, et seq.

Migratory Bird Treaty Act. 16 USC 703 et seq.

Rivers and Harbors Appropriations Act of 1899. 33 USC 403.

U.S. Environmental Protection Agency (EPA). 1993. "Chapter 7: Management Measures for Wetlands, Riparian Areas, and Vegetated Treatment Systems, Glossary." In *Guidance Specifying Management Measures for Sources of Nonpoint Pollution in Coastal Waters.* EPA 840-B-92-002 (January 1993). Office of Wetlands, Oceans & Watersheds. Available at http://www.epa.gov/owow/nps/MMGI/. Accessed on January 15, 2013.

U.S. Nuclear Regulatory Commission (NRC). 2012. *Terrestrial Environmental Studies for Nuclear Power Stations.* Regulatory Guide 4.11, Rev. 2, Washington, D.C. March.

U.S. Nuclear Regulatory Commission (NRC). 2013. *Generic Environmental Impact Statement for License Renewal of Nuclear Plants.* NUREG-1437, Vols. 1, 2, and 3, Revision 1. Office of Nuclear Reactor Regulation, Washington, D.C.

U.S. NUCLEAR REGULATORY COMMISSION

ENVIRONMENTAL STANDARD
REVIEW PLAN

OFFICE OF NUCLEAR REACTOR REGULATION

3.6 HISTORIC AND CULTURAL RESOURCES

I. AREAS OF REVIEW

This environmental standard review plan (ESRP) provides guidance for identifying and describing historic and cultural resources that may be impacted by license renewal. Historic and cultural resources include prehistoric and historic era archaeological sites, historic structures, and traditional cultural properties. To identify historic and cultural resources at a nuclear plant site, the staff should review historic building surveys, archaeological surveys, historical research, surveys of cultural groups with a history of using the area in and around the plant site, and consultations with State Historic Preservation Office [SHPO], Tribal Historic Preservation Office [THPO], Office of the State Archaeologist, American Indian Tribes, and other knowledgeable groups and agencies. Included in the identification process is an assessment of historic and cultural resources, as determined by criteria in the National Historic Preservation Act of 1966 (NHPA), the National Environmental Policy Act of 1969 (NEPA), and in consultation with affected Indian Tribes. The descriptions of the historic and cultural resources should be of sufficient detail to permit subsequent staff assessment and evaluation of specific impacts of the proposed action.

The scope should include (1) an analysis of the effects of continued plant operations during the license renewal term and refurbishment on historic and cultural resources within the area of potential effects (APE) in sufficient detail to allow the reviewer to predict the potential for impacts on these resources and evaluate the significance of impacts and (2) preparation of input to the supplemental environmental impact statement (SEIS).

Historic properties means any prehistoric or historic districts, sites, buildings, structures, or objects included in, or eligible for inclusion in, the *National Register of Historic Places* (NRHP) maintained by the Secretary of the Interior (36 CFR 60). This term includes properties of traditional religious and cultural importance to an Indian tribe or Native Hawaiian organization and that meet the NRHP criteria (36 CFR 63). The term also includes archaeological resources, such as artifacts, records, and remains, that are related to and located within such prehistoric or historic districts, sites, buildings, or structures.

Data and Information Needs

The type of data and information needed would be affected by nuclear power plant site- and plant-specific factors, the amount of previous survey work conducted in the area; and the level of interest from agencies, Tribes, and other consulting parties. The following data or information should be included in this section, and should be obtained from the applicant's environmental report (ER), as well as other information collected during verification and assessment:

- a summary of the cultural history of the area, from the beginning of human settlement to the early 20th century

- a summary of the land use history at the plant site documenting past levels of ground disturbance

- a description of the cultural resource sensitivity within a 10-mile radius of the plant boundaries

- a copy of the site map that identifies the APE

- a summary of previous cultural resource investigations conducted onsite or within the immediate area and the types of resources that have been discovered during previous research

- a description of the historic and cultural resources identified within the APE that are in or eligible for inclusion in the NRHP or are included in State or local registers or inventories of historic and cultural resources or are unevaluated (from the ER and consultation with Federal, State, regional, local, and affected American Indian Tribes)

- a description of the information from the applicant's reconnaissance or pedestrian surveys, and consultation efforts with the SHPO, appropriate American Indian Tribes, or members of the public the applicant used to assess historic and cultural resources within the APE

- information related to past evaluations of known historic and cultural resources per significance criteria for eligibility for the NRHP (36 CFR 60), and associated consultations with the SHPO or THPO, local preservation officials, or American Indian Tribal officials

- information on how the plant considers historic properties during operations

II. ACCEPTANCE CRITERIA

Acceptance criteria for the review of historic and cultural resources impacts during the renewal term are based on the relevant requirements of the following:

- 10 CFR 51.45(d), "Status of Compliance." The environmental report shall list all Federal permits, licenses, approvals and other entitlements which must be obtained in connection with the proposed action and shall describe the status of compliance with these requirements. The environmental report shall also include a discussion of the status of compliance with applicable environmental quality standards and requirements including, but not limited to, applicable zoning and land-use regulations, and thermal and other water pollution limitations or requirements which have been imposed by Federal, State, regional, and local agencies having responsibility for environmental protection.

- 10 CFR 51.53(c)(2). The report must contain a description of the proposed action, including the applicant's plans to modify the facility or its administrative control procedures as described in accordance with 10 CFR 54.21 of this chapter. This report must describe in detail the affected environment around the plant, the modifications directly affecting the environment or any plant effluents, and any planned refurbishment activities. In addition, the applicant shall discuss in this report the environmental impacts of alternatives and any other matters discussed in 10 CFR 51.45.

- 10 CFR 51.53(c)(3)(ii)(K). All applicants shall identify any potentially affected historic or archaeological properties and assess whether any of these properties will be affected by future plant operations and any planned refurbishment activities in accordance with the National Historic Preservation Act.

- 10 CFR 51.70(b). The draft environmental impact statement will be concise, clear, and analytic, and written in plain language with appropriate graphics.…The format provided in Section 1(a) of Appendix A of this subpart should be used. The NRC staff will independently evaluate and be responsible for the reliability of all information used in the draft environmental impact statement.

- 10 CFR Part 51, Appendix A to Subpart A, para. 6, concerning affected environment. The environmental impact statement will succinctly describe the environment to be affected by the proposed action. Data and analyses in the statement will be commensurate with the importance of the impact, with less important material summarized, consolidated, or simply referenced. Effort and attention will be concentrated on important issues; useless bulk will be eliminated.

- 36 CFR Part 800 defines the process by which a Federal agency meets the requirements under Sections 106 and 110 of the NHPA to ensure that agency-assisted or agency-licensed undertakings consider the effects of the undertaking on historic properties included in or eligible for the NRHP. Under this regulation, the NRC is required to identify and evaluate all historic properties in the project area and take measures to mitigate adverse effects. As indicated in 36 CFR 800.8(c), Section 106 can be integrated with NEPA reviews.

- 36 CFR Part 63 contains guidance by which historic properties are evaluated and determined eligible for listing on the NRHP.

Technical Rationale

The review conducted under this ESRP leads to preparation of a section describing the affected environment for the SEIS that provides background information to be used in evaluating historic and cultural resource impacts of continued plant operations and refurbishment activities associated with license renewal.

III. REVIEW PROCEDURE

The description of historic and cultural resources should form the basis for the impact assessment review described by ESRP Section 4.6 to establish the historical, cultural, and archaeological characteristics that are most likely to be affected by license renewal. The following review steps are suggested:

1. Review the discussion of potential impacts from continued plant operations and refurbishment activities associated with license renewal on historic and cultural resources in the *Generic Environmental Impact Statement for License Renewal of Nuclear Power Plant* (GEIS; NUREG-1437) to identify the information considered and the conclusions reached. This step establishes the base for evaluation of information identified by the applicant, the public, and the staff.

2. Review the historic and cultural resource identification and impact assessment prepared by the applicant in the ER. Develop a list of questions that arise during the review of the ER.

3. Coordinate with the environmental project manager (EPM) to identify consulting parties, such as SHPO/THPO, Advisory Council on Historic Preservation, appropriate Indian Tribes, and any other group or organization that has been identified; as the review progresses, additional letters could be sent to other interested groups.

4. Contact the appropriate SHPO/THPO and arrange for a meeting during the site audit to explain the proposed action, the NRC's approach to conducting the cultural resource assessment (Section 106 integration with NEPA), and opportunities for the SHPO/THPO to comment and to capture any issues, concerns, and expectations. If the SHPO/THPO has comments or information that add to or amplify that which was provided by the applicant, request that the SHPO/THPO forward, by letter to the staff, these additional comments. The SHPO can alert the staff to relevant state and local laws, orders, ordinances, or regulations aimed at the preservation of cultural resources applicable to the proposed action. Be sure to discuss the following:

   - the APE

   - the status of all historic properties within the APE

   - the data necessary to accomplish the impact assessment

- additional organizations or individuals that might be able to assist in identifying and locating historic and cultural resources

- guidance on consulting with the appropriate Indian Tribes

5. Contact the State Archaeological Site Files Office and the State Historic Buildings Survey Office and arrange to review the files for any recorded sites or buildings in and adjacent to the APE, and information concerning the cultural resource surveys that have been conducted. In many cases, the site files are located at the SHPO's office. The sites and buildings identified should match the findings presented in the ER.

6. Review comments received during the scoping process to identify any issues associated with historic and cultural resources.

7. Collect additional information during the site audit. Review cultural resource assessments prepared by the applicant and those prepared previously. Tour the APE to see locations where pre-construction and construction activities would occur; visit any known sites within the APE.

8. Compare the information provided by the applicant with that obtained from the SHPO/THPO and the NRHP and resolve any differences in identification, NRHP status, and location of historic and cultural resources within the APE.

9. Compare the areas of the plant site potentially affected by refurbishment activities with known cultural resource surveys to determine if the area has been surveyed before and whether there is potential for important cultural resources to be located there. Factors to consider include whether areas have been heavily disturbed in the past, availability of water or other critical resources, and other environmental characteristics, such as slope.

10. Document the findings in the SEIS. Ensure that any sensitive archaeological reports are withheld from public disclosure per Section 304 of the NHPA or have been redacted (e.g., the location of historic properties is considered protected information). All redactions must be approved by the SHPO.

11. Document in the SEIS that the NRC is using NEPA to fulfill Section 106 responsibilities (36 CFR 800.8).

IV. EVALUATION FINDINGS

The amount of information presented in the SEIS would be governed by the extent and significance of the historic properties present in the APE and the effects of continued plant operations, refurbishment, and decommissioning activities on historic and cultural resources. The reviewer should verify that historic and cultural resources have been identified and described in sufficient detail to provide the basis for subsequent analysis and assessment of these impacts.

V. IMPLEMENTATION

The method described in this ESRP would be used by the staff in evaluating conformance with the Commission's regulations, except in those cases in which the applicant for license renewal proposes an acceptable alternative for complying with specified portions of the regulations.

VI. BIBLIOGRAPHY

10 CFR Part 51. *Code of Federal Regulations*, Title 10, *Energy*, Part 51, "Environmental Protection Regulations for Domestic Licensing and Related Regulatory Functions."

36 CFR Part 60. *Code of Federal Regulations*, Title 36, *Parks, Forests, and Public Property*, Part 60, "National Register of Historic Places."

36 CFR Part 63. *Code of Federal Regulations*, Title 36, *Parks, Forests, and Public Property*, Part 63, "Determinations of Eligibility for Inclusion in the National Register of Historic Places."

36 CFR Part 800. *Code of Federal Regulations*, Title 36, *Parks, Forests, and Public Property*, Part 800, "Protection of Historic Properties."

48 FR 44716. U.S. Department of the Interior (DOI). Archeology and Historic Preservation; Secretary of the Interior's Standards and Guidelines. 1983.

National Environmental Policy Act of 1969. 42 USC 4321 et seq.

National Historic Preservation Act of 1966. 16 USC 470 et seq.

U.S. National Park Service (NPS). 1985. "Guidelines for Local Surveys: A Basis for Preservation Planning." *National Register Bulletin* No. 24. U.S. Department of the Interior, Washington, D.C.

U.S. National Park Service (NPS). 1990. "Guidelines for Evaluating and Documenting Traditional Cultural Properties." *National Register Bulletin* No. 38. U.S. Department of the Interior, Washington, D.C.

U.S. National Park Service (NPS). 1991. "How to Apply the National Register Criteria for Evaluation." *National Register Bulletin* No. 15. U.S. Department of the Interior, Washington, D.C.

U.S. Nuclear Regulatory Commission (NRC). 2013a. *Preparation of Environmental Reports for Nuclear Power Plant License Renewal Applications*. Regulatory Guide 4.2, Supplement 1, Revision 1. Washington, D.C.

U.S. Nuclear Regulatory Commission (NRC). 2013b. *Generic Environmental Impact Statement for License Renewal of Nuclear Plants*. NUREG-1437, Vols. 1, 2, and 3, Revision 1. Office of Nuclear Reactor Regulation, Washington, D.C.

U.S. NUCLEAR REGULATORY COMMISSION
ENVIRONMENTAL STANDARD
REVIEW PLAN
OFFICE OF NUCLEAR REACTOR REGULATION

3.7 SOCIOECONOMICS

I. AREAS OF REVIEW

This environmental standard review plan (ESRP) guides the review and consideration of socioeconomic factors that could be directly or indirectly affected by changes in nuclear power plant operations. A nuclear power plant and the communities that support it can be described as a dynamic socioeconomic system. The communities provide the people, goods, and services required to operate the nuclear power plant. Power plant operations, in turn, provide wages and benefits for people and dollar expenditures for goods and services. Payments for these goods and services create additional employment and income opportunities in the community. The measure of a community's ability to support the operational demands of a nuclear power plant depends on the ability of the community to respond to changing environmental, social, economic, and demographic conditions.

The socioeconomic region of influence (ROI) is defined by the areas where plant employees and their families reside, spend their income, and use their benefits, thereby affecting the economic conditions of the region. The following sections describe the housing, public services, offsite land use, visual aesthetics and noise, population demography, and the economy in the ROI surrounding the plant site.

The scope should include the establishment of the nature and extent of present and planned socioeconomic factors within these areas that might be impacted or modified as a result of continued plant operations and refurbishment associated with license renewal. Descriptions of minority and low-income populations, their locations, practices, and customs would have to be addressed in this ESRP. Those features of the socioeconomic environment that are not expected to result in socioeconomic impacts should be discussed briefly and in a summary manner.

USNRC ENVIRONMENTAL STANDARD REVIEW PLAN

Data and Information Needs

The reviewer should consult the *Generic Environmental Impact Statement for License Renewal of Nuclear Plants* (GEIS; NUREG-1437, Volumes 1, 2, and 3, Revision 1), before undertaking extensive data collection since socioeconomic impacts may not require an extensive data collection effort.

The types of socioeconomic data and information needed for analysis would depend on the nuclear power plant site- and plant-specific factors. The following data or information may be needed:

- most recent average annual total number of permanent plant workers and county of residence, average number of plant outage workers, frequency, and duration (in days or weeks)

- U.S. Bureau of Census information and data related to the ROI (by county) economic base, including:

  - housing: total number of units, number of occupied units, number of vacant units, vacancy rate, and median value

  - demographic information by race and ethnicity and population growth forecasts by county

  - transient (seasonal) population including students attending colleges and universities within 50 miles of the plant

  - civilian labor force by county

  - largest industrial employment by industrial sector category (North American Industry Classification System [NAICS] code)

  - median household income and per capita income

  - percent of families and individuals living below the Census poverty threshold

  - unemployment

- public water supply system information by source (groundwater or surface water, average daily production, system design capacity, and population served)

- information about the local public schools: school district(s), total enrollment

- information on local transportation systems: condition of site access roads, average annual daily traffic volume and road capacity

- information on offsite land use

- visual description of the plant

- Census of Agriculture (U.S. Department of Agriculture) information on migrant farm labor in the ROI (by county), including:

- number of farms and farm workers working less than 150 days
- number of farms reporting migrant farm labor
- number of farms with hired farm labor

- list of major employers in ROI

- tax payment information including local tax authorities (i.e., county, municipality, and public school district) directly affected by plant operations

- public recreational facilities, including present and projected capacity and percentage of utilization

II. ACCEPTANCE CRITERIA

Acceptance criteria for evaluating community characteristics are based on meeting the following regulations:

- 10 CFR 51.45(d), "Status of Compliance." The environmental report shall list all Federal permits, licenses, approvals and other entitlements which must be obtained in connection with the proposed action and shall describe the status of compliance with these requirements. The environmental report shall also include a discussion of the status of compliance with applicable environmental quality standards and requirements including, but not limited to, applicable zoning and land-use regulations, and thermal and other water pollution limitations or requirements which have been imposed by Federal, State, regional, and local agencies having responsibility for environmental protection.

- 10 CFR 51.53(c)(2). The report must contain a description of the proposed action, including the applicant's plans to modify the facility or its administrative control procedures as described in accordance with 10 CFR 54.21 of this chapter. This report must describe in detail the affected environment around the plant, the modifications directly affecting the environment or any plant effluents, and any planned refurbishment activities. In addition, the applicant shall discuss in this report the environmental impacts of alternatives and any other matters discussed in 10 CFR 51.45.

- 10 CFR 51.70(b). The draft environmental impact statement will be concise, clear, and analytic, and written in plain language with appropriate graphics.…The format provided in Section 1(a) of Appendix A of this subpart should be used. The Nuclear Regulatory Commission (NRC) staff will independently evaluate and be responsible for the reliability of all information used in the draft environmental impact statement.

- 10 CFR Part 51, Appendix A to Subpart A, para. 6, concerning affected environment. The environmental impact statement will succinctly describe the environment to be affected by the proposed action. Data and analyses in the statement will be commensurate with the importance of the impact, with less important material summarized, consolidated, or simply referenced. Effort and attention will be concentrated on important issues; useless bulk will be eliminated.

Technical Rationale

The review conducted under this ESRP leads to preparation of a section describing the affected environment for the SEIS that provides background information to be used in evaluating socioeconomic impacts of continued plant operations and refurbishment activities associated with license renewal.

III. REVIEW PROCEDURE

The reviewer's analysis of socioeconomic factors should be closely linked with the impact-assessment review described in ESRP Chapter 4.7 to establish the socioeconomic factors most likely to be affected by license renewal. The following review steps are suggested:

1. Review the socioeconomics discussion in the GEIS to identify the information considered and the conclusions reached. This step establishes the basis for evaluation of information identified by the applicant and the public.

2. Determine if there is new information that should be evaluated. The following sources of information should be included in the search for new information:

    - The applicant's ER. In reviewing the applicant's ER, consider any new information.
    - Records of public scoping meetings and correspondence related to the application. Compare information presented by the public with information considered in the GEIS.

3. Describe the socioeconomic characteristics of the counties within the ROI.

4. Address the following factors in the screening process to identify population influx:

    - labor force
    - transportation
    - housing availability
    - public services
    - regional economy

5. Describe potentially impacted areas of the ROI and their associated communities in the following terms (the extent and detail of the descriptions should be in proportion to the magnitude of the impacts anticipated, and only those terms necessary for subsequent impact evaluation should be used):

    - demography
    - housing
    - economic base

- social services and public facilities
- local transportation
- public water supply
- education
- recreation
- taxation

6. The reviewer should ensure that the socioeconomic information forms a basis for an assessment of impacts in the supplemental environmental impact statement (SEIS).

7. Based on the results of the assessments listed above, prepare the following for the SEIS:

   - a brief description of the workforce required during continued plant operations during the license renewal term and refurbishment
   - a discussion (qualitative or quantitative, as appropriate) of the local population(s) and public infrastructure and services

IV. EVALUATION FINDINGS

The amount of information in the SEIS would be governed by the extent and significance of the potential effects of continued plant operations and refurbishment associated with license renewal on public services.

V. IMPLEMENTATION

The method described in this ESRP would be used by the staff in evaluating conformance with the Commission's regulations, except in those cases in which the applicant for license renewal proposes an acceptable alternative for complying with specified portions of the regulations.

VI. BIBLIOGRAPHY

10 CFR Part 51. *Code of Federal Regulations*, Title 10, *Energy*, Part 51, "Environmental Protection Regulations for Domestic Licensing and Related Regulatory Functions."

U.S. Nuclear Regulatory Commission (NRC). 2013. *Generic Environmental Impact Statement for License Renewal of Nuclear Plants* (GEIS). NUREG-1437, Vols. 1, 2, and 3, Revision 1. Office of Nuclear Reactor Regulation, Washington, D.C.

U.S. NUCLEAR REGULATORY COMMISSION

ENVIRONMENTAL STANDARD REVIEW PLAN

OFFICE OF NUCLEAR REACTOR REGULATION

3.8 HUMAN HEALTH

I. AREAS OF REVIEW

This environmental standard review plan (ESRP) provides guidance for the discussion of radiological impacts of nuclear power plants. The scope includes descriptions of the radioactive waste management program, radiological environmental monitoring and radioactive effluent release programs, and occupational radiation exposure.

Data and Information Needs

The types of data and information needed would be affected by nuclear power plant site- and plant-specific factors. The following data or information may be needed:

- a description of the radioactive liquid, gaseous, and solid waste management and effluent control systems and information on effluents released into the environment and waste stored onsite

- historical data on occupational doses to plant workers (from NUREG-0713, "Occupational Radiation Exposure at Commercial Nuclear Power Reactors and Other Facilities")

- a description of the radiological environmental monitoring program and environmental data (from the applicant's annual environmental operating reports)

- historical maximum doses to a member of the public (from the applicant's annual radioactive effluent release reports)

- information on the potential changes in radiological impacts from continued plant operations during the renewal term

- information on the radiological impacts of refurbishment

II. ACCEPTANCE CRITERIA

The reviewer should ensure that the introductory and descriptive paragraphs prepared under this ESRP are consistent with the following regulations:

- 10 CFR 51.45(d), "Status of Compliance." The environmental report shall list all Federal permits, licenses, approvals and other entitlements which must be obtained in connection with the proposed action and shall describe the status of compliance with these requirements. The environmental report shall also include a discussion of the status of compliance with applicable environmental quality standards and requirements including, but not limited to, applicable zoning and land-use regulations, and thermal and other water pollution limitations or requirements which have been imposed by Federal, State, regional, and local agencies having responsibility for environmental protection.

- 10 CFR 51.53(c)(2). The report must contain a description of the proposed action, including the applicant's plans to modify the facility or its administrative control procedures as described in accordance with 10 CFR 54.21 of this chapter. This report must describe in detail the affected environment around the plant, the modifications directly affecting the environment or any plant effluents, and any planned refurbishment activities. In addition, the applicant shall discuss in this report the environmental impacts of alternatives and any other matters discussed in 10 CFR 51.45.

- 10 CFR 51.53(c)(3)(ii)(G). If the applicant's plant uses a cooling pond, lake, or canal or discharges into a river, an assessment of the impact of the proposed action on public health from thermophilic organisms in the affected water must be provided.

- 10 CFR 51.53(c)(3)(ii)(H). If the applicant's transmission lines that were constructed for the specific purpose of connecting the plant to the transmission system do not meet the recommendations of the National Electric Safety Code for preventing electric shock from induced currents, an assessment of the impact of the proposed action on the potential shock hazard from the transmission lines must be provided.

- 10 CFR 51.70(b). The draft environmental impact statement will be concise, clear, and analytic, and written in plain language with appropriate graphics....The format provided in Section 1(a) of Appendix A of this subpart should be used. The NRC staff will independently evaluate and be responsible for the reliability of all information used in the draft environmental impact statement.

- 10 CFR Part 51, Appendix A to Subpart A, para. 6, concerning affected environment. The environmental impact statement will succinctly describe the environment to be affected by the proposed action. Data and analyses in the statement will be commensurate with the importance of the impact, with less important material summarized, consolidated, or simply referenced. Effort and attention will be concentrated on important issues; useless bulk will be eliminated.

<u>Technical Rationale</u>

The review conducted under this ESRP leads to preparation of a section describing the affected environment for the supplemental environmental impact statement (SEIS) that provides background information to be used in evaluating human health impacts of continued plant operations and refurbishment activities associated with license renewal.

III. REVIEW PROCEDURES

The SEIS section to be prepared on the radiological impacts is informational in nature. No specific analysis is required. The following review steps are suggested:

1. Review the discussion of Human Health in Section 3.9 in the *Generic Environmental Impact Statement for License Renewal of Nuclear Power Plant* (GEIS; NUREG-1437).

2. Obtain historic information (typically 5 years of data) on radioactive effluents released from the applicant's plant.

3. Obtain information on expected radioactive releases and exposures from refurbishment activities, if any.

4. Obtain information on projected changes in radioactive releases and exposures from operations during the renewal term, if any.

5. Obtain historical information (typically 5 years of data) on the radiological environmental monitoring program.

6. Obtain historical information (typically 5 years of data) on the occupational doses to plant workers.

7. Prepare a section describing the radiological programs and systems for the SEIS. This section should include summary descriptions of the applicant's radioactive effluent monitoring and radiological environmental monitoring programs. It should also include a discussion of doses received by members of the public and plant workers for the most recent calendar year and the trend of such doses for the most recent 5 years of plant operation. Doses should be compared with relevant regulatory requirements, e.g., Appendix I to 10 CFR 50, 10 CFR 20.1201, and 10 CFR 20.1301. For the radiological environmental monitoring program, provide a summary of the results for the most recent calendar year and a trend of the data for the most recent 5 years of plant operation.

IV. EVALUATION FINDINGS

The level of detail of SEIS input would depend on plant- and site-specific factors. The information included in the SEIS should be scaled according to the anticipated magnitudes of the expected impacts. The reviewer should verify that the radiological impact descriptions are consistent, accurate, and given in sufficient detail to serve the needs of the reviewers for ESRPs in other chapters.

V. IMPLEMENTATION

The method described in this ESRP would be used by the staff in evaluating conformance with the Commission's regulations, except in those cases in which the applicant for license renewal proposes an acceptable alternative for complying with specified portions of the regulations.

VI. BIBLIOGRAPHY

10 CFR Part 20. *Code of Federal Regulations*, Title 10, *Energy,* Part 20, "Standards for Protection against Radiation."

10 CFR Part 50. *Code of Federal Regulations*, Title 10, *Energy,* Part 50, "Domestic Licensing of Production and Utilization Facilities."

10 CFR Part 51. *Code of Federal Regulations*, Title 10, *Energy,* Part 51, "Environmental Protection Regulations for Domestic Licensing and Related Regulatory Functions."

U.S. Nuclear Regulatory Commission (NRC). 2013. *Generic Environmental Impact Statement for License Renewal of Nuclear Plants.* NUREG-1437, Vols. 1, 2, and 3, Revision 1. Office of Nuclear Reactor Regulation, Washington, D.C.

3.9 ENVIRONMENTAL JUSTICE

I. AREAS OF REVIEW

This environmental standard review plan (ESRP) provides guidance for the preparation of a supplemental environmental impact statement (SEIS) section describing the staff's identification and description of low-income and minority populations that could be disproportionately impacted by continued operations and refurbishment activities associated with license renewal.

The scope includes consideration and discussion of methods that are used to identify and locate minority and low-income populations, the location and significance of any populations that are particularly sensitive, and any additional information pertaining to any identified disproportional preconditions or sensitivities of minority and low-income populations that could be disproportionately high and adversely impacted by continued plant operation and refurbishment. The descriptions to be provided by this review should be of sufficient detail to permit subsequent staff assessment and evaluation of specific impacts as provided in ESRP 4.9.

Data and Information Needs

The type of data and information needed will be affected by site- and station-specific factors, and the degree of detail should be modified according to the anticipated magnitude of the potential impacts. The information needed can usually be obtained from the applicant's environmental report (ER) and from other sources as discussed below. The following data or information should be obtained:

- The location of minority and low-income populations within 50-miles (80-kilometers) of the nuclear power plant site, including offsite areas that could be impacted by continued reactor operations during the license renewal term and refurbishment activities associated with license renewal (from the ER, public contacts, and consultations with Federal, State, regional, local, and representative s of affected Native American Tribes). Demographic data would be available from EJView, an online geographic information system (GIS) tool offered by the U.S. Environmental Protection Agency, and the Bureau of the Census block data and Topologically Integrated Geographic Encoding and Referencing (TIGER) geographic system files.[2]

- Comments and concerns expressed by representatives of minority and low-income communities located near the nuclear power plant site (from the environmental report [ER] and comments made during scoping meetings and other contacts with the public). As part of scoping, it is important to interview representatives of minority communities and other regional contacts (such as social service agencies) having specific knowledge about the locations, resource dependencies, customs and practices, and preexisting health and socioeconomic conditions of minority and low-income populations in the region. This will assist the Nuclear Regulatory Commission (NRC) in ensuring that minority and low-income communities, including transient populations, affected by the proposed action are not overlooked and in assessing the potential for significant impacts unique to those communities. The resources devoted to this specific outreach should be a matter of professional judgment and should be commensurate with the likelihood of disproportionately high and adverse impacts on minority and low-income populations. Both the outreach process and results of the interviews, especially information about circumstances that could lead to disproportionately high and adverse impacts from the proposed action, should be documented in the SEIS.

- A description of resources, customs and practices, circumstances of living (e.g., migrant labor), or preconditions (e.g., preexisting health conditions or access to particular facilities or locations) of particular minority or low-income populations that may make them likely to experience disproportionately high and adverse environmental impacts from the proposed action. If there are no such populations or mechanisms, the section should describe the search process and state that none was identified.

II. ACCEPTANCE CRITERIA

The acceptance criteria for environmental justice information are based on the relevant requirements of the following:

- Executive Order 12898 (59 FR 7629) provides guidance concerning Federal actions to address environmental justice in minority and low-income populations.

- "Policy Statement on the Treatment of Environmental Justice Matters in NRC Regulatory and Licensing Actions," (69 FR 52040) affirms that the NRC is committed to the general goals of Executive Order 12898 and states that the NRC strives to meet those goals as part of its NEPA review process for licensing actions.

---

[2]   The TIGER system is accessible online at the Census Bureau TIGER website http://www.census.gov/geo/www/tiger/.

- 10 CFR 51.53(c)(2). The report must contain a description of the proposed action, including the applicant's plans to modify the facility or its administrative control procedures as described in accordance with 10 CFR 54.21 of this chapter. This report must describe in detail the affected environment around the plant, the modifications directly affecting the environment or any plant effluents, and any planned refurbishment activities. In addition, the applicant shall discuss in this report the environmental impacts of alternatives and any other matters discussed in 10 CFR 51.45.

- 10 CFR 51.53(c)(3) (ii)(N). Applicants shall provide information on the general demographic composition of minority and low-income populations and communities (by race and ethnicity) residing in the immediate vicinity of the plant that could be affected by the renewal of the plant's operating license, including any planned refurbishment activities, and ongoing and future plant operations.

- 10 CFR 51.70(b). The draft environmental impact statement will be concise, clear, and analytic, and written in plain language with appropriate graphics....The format provided in Section 1(a) of Appendix A of this subpart should be used. The NRC staff will independently evaluate and be responsible for the reliability of all information used in the draft environmental impact statement.

Additional regulatory positions and specific criteria in support of the regulations identified above are as follows:

- Council on Environmental Quality (CEQ) guidance for addressing environmental justice, *Environmental Justice: Guidance under the National Environmental Policy Act*, December 10, 1997 (CEQ 1997).

- Guidelines for specific information requirements for environmental justice determinations are described in Office of Nuclear Reactor Regulation (NRR) Office Instruction LIC-203, Revision 2: *Procedural Guidance for Preparing Environmental Assessments and Considering Environmental Issues*. NRR Office Instruction LIC-203 is revised periodically. Obtain a copy of the latest revision for current guidance.

- 10 CFR Part 51, Appendix A to Subpart A, para. 6, concerning affected environment. The environmental impact statement will succinctly describe the environment to be affected by the proposed action. Data and analyses in the statement will be commensurate with the importance of the impact, with less important material summarized, consolidated, or simply referenced. Effort and attention will be concentrated on important issues; useless bulk will be eliminated.

Technical Rationale

The review conducted under this ESRP leads to preparation of a section describing the affected environment for the SEIS that provides background information to be used in evaluating the environmental justice impacts from continued plant operations and refurbishment activities associated with license renewal.

III. REVIEW PROCEDURES

The review procedure should be as follows:

1. Identify minority and low-income populations within a 50-mile (80-kilometer) radius of the nuclear plant. For each census block group within this area, minority and low-income populations are identified when (1) the minority or low-income population of an impacted area exceeds 50 percent or (2) the minority or low-income population percentage of the impacted area is meaningfully greater than the minority or low-income population percentage in the general population or other appropriate unit of geographic analysis (e.g., 50-mile radius geographic area or county) All block groups with minority and low-income percentages higher than the geographic area should be identified on 50-mile radius maps.

2. Identify environmental justice issues and unique characteristics of minority and low-income populations/communities during the scoping process.

   - Determine geographic distribution by race, ethnicity, and poverty, as well as delineation of Tribal lands. Identify any unique characteristics of minority and low-income populations and the "special character" of communities located near the nuclear plant.

     In calculating the minority populations, individual(s) who are members of the following population groups Census are considered minority individuals:

     - Race: (Not Hispanic or Latino)
       Black or African American
       American Indian or Alaska Native

Asian
Native Hawaiian and Other Pacific Islander
Some other race
Two or more races

- Ethnicity:

Hispanic or Latino (of any race)

Low-income population is defined as individuals or families living below the poverty level as defined by the U.S. Census Bureau (e.g., the U.S. Census Bureau's Current Population Reports, Series P-60 on Income and Poverty).

- Sources of information for determining geographic distribution and location of minority populations:

  - EJView online geographic information system (GIS) tool offered by the U.S. Environmental Protection Agency.

  - Local governments

  - State agencies

  - Local universities

3. The reviewer's analysis of minority and low-income populations should be closely linked with the impact-assessment review of environmental issues described by the ESRPs 3.1 through 3.8 and 4.1 through 4.8 to establish the environmental pathways by which minority and low-income households are most likely to be affected in a disproportionately high and adverse manner, if any.

3.9-5

IV. EVALUATION FINDINGS

The depth and extent of the input to the SEIS would depend on plant- and site-specific factors. The level of detail of information included in the SEIS should be scaled according to the anticipated magnitudes of the expected impacts on populations living in the vicinity of the nuclear power plant in general and on the identified minority and low-income populations in particular from the continued operation and refurbishment during the license renewal term. The following information should be included in the SEIS:

- A general description of the location of minority and low-income populations within the region surrounding the site. This description should ordinarily be accompanied by two maps that highlight the location of minority and low-income populations, respectively. These maps would ordinarily be based on the most recent Census of Population, supplemented by other information, if available.

- A description of affected minority and low-income populations such as minority communities exceptionally dependent on subsistence resources or identifiable in compact locations, such as a Native American settlement.

- A brief description of any additional important cultural, economic, or human health facts that may result in disproportionately high and adverse environmental (including socioeconomic) impacts.

- A brief description of the overall results and adequacy of any surveys (archival or field) that were conducted by the applicant.

V. IMPLEMENTATION

The method described in this ESRP would be used by the staff in evaluating conformance with the Commission's regulations, except in those cases in which the applicant for license renewal proposes an acceptable alternative for complying with specified portions of the regulations.

VI. BIBLIOGRAPHY

10 CFR Part 51. *Code of Federal Regulations*, Title 10, *Energy,* Part 51, "Environmental Protection Regulations for Domestic Licensing and Related Regulatory Functions."

69 FR 52040. U.S. Nuclear Regulatory Commission. Policy Statement on the Treatment of Environmental Justice Matters in NRC Regulatory and Licensing Actions. August 24, 2004.

Council on Environmental Quality (CEQ). *Environmental Justice: Guidance under the National Environmental Policy Act.* December 10, 1997

Executive Order 12898. *Federal Actions to Address Environmental Justice in Minority Populations and Low-Income Populations.* 59 FR 7629. February 16, 1994.

NUREG-1555, Supplement 1

U.S. NUCLEAR REGULATORY COMMISSION

ENVIRONMENTAL STANDARD REVIEW PLAN

OFFICE OF NUCLEAR REACTOR REGULATION

U.S. Nuclear Regulatory Commission (NRC). Appendix D – Environmental Justice Guidance and Flow Chart. In *Procedural Guidance for Preparing Environmental Assessments and Considering Environmental Issues*. Office of Nuclear Reactor Regulation Instruction Change Notice, LIC-203, Revision 1, Washington D.C. May 24, 2004.

U.S. Nuclear Regulatory Commission (NRC). 2009. *Procedural Guidance for Preparing Environmental Assessments and Considering Environmental Issues*. Office of Nuclear Reactor Regulation Instruction Change Notice, LIC-203, Revision 2, February 17, 2009, Washington, D.C.

U.S. Nuclear Regulatory Commission (NRC). 2013. *Generic Environmental Impact Statement for License Renewal of Nuclear Plants*. NUREG-1437, Vols. 1, 2, and 3, Revision 1. Office of Nuclear Reactor Regulation, Washington, D.C.

3.9-7

U.S. NUCLEAR REGULATORY COMMISSION
ENVIRONMENTAL STANDARD
REVIEW PLAN
OFFICE OF NUCLEAR REACTOR REGULATION

3.10 WASTE MANAGEMENT

3.10.1 RADIOACTIVE WASTE SYSTEMS

I. AREAS OF REVIEW

This environmental standard review plan (ESRP) provides guidance for the preparation of a supplemental environmental impact statement (SEIS) section describing the applicant's waste management and effluent control systems.

The scope includes describing the existing systems, describing any changes to the systems to be made during the license renewal term or refurbishment, and listing the effluent release points.

<u>Data and Information Needs</u>

The types of data and information needed would be affected by nuclear power plant site- and plant-specific factors; the level of detail should be scaled according to the anticipated magnitude of the potential impacts. The following data or information may be needed:

*   a description of the radioactive liquid and gaseous waste management and effluent control systems

*   identification of sources of radioactive liquid and gaseous waste material within the plant

*   identification of principal release points for radioactive materials to the environment and historical information on composition of discharges

3.10.1-1

- identification of any onsite direct radiation sources outside of the plant (e.g., storage of contaminated equipment, low-level radioactive waste storage, or storage of used steam generators)

- information on the changes in radiological waste impacts from operation that are expected during the renewal term

II. ACCEPTANCE CRITERIA

The reviewer should ensure that the introductory and descriptive paragraphs prepared under this ESRP are consistent with the following regulation:

- 10 CFR 51.45(d), "Status of Compliance." The environmental report shall list all Federal permits, licenses, approvals and other entitlements which must be obtained in connection with the proposed action and shall describe the status of compliance with these requirements. The environmental report shall also include a discussion of the status of compliance with applicable environmental quality standards and requirements including, but not limited to, applicable zoning and land-use regulations, and thermal and other water pollution limitations or requirements which have been imposed by Federal, State, regional, and local agencies having responsibility for environmental protection.

- 10 CFR 51.53(c)(2). The report must contain a description of the proposed action, including the applicant's plans to modify the facility or its administrative control procedures as described in accordance with 10 CFR 54.21 of this chapter. This report must describe in detail the affected environment around the plant, the modifications directly affecting the environment or any plant effluents, and any planned refurbishment activities. In addition, the applicant shall discuss in this report the environmental impacts of alternatives and any other matters discussed in 10 CFR 51.45.

- 10 CFR 51.70(b). The draft environmental impact statement will be concise, clear, and analytic, and written in plain language with appropriate graphics.... The format provided in Section 1(a) of Appendix A of this subpart should be used. The NRC staff will independently evaluate and be responsible for the reliability of all information used in the draft environmental impact statement.

- 10 CFR Part 51, Appendix A to Subpart A, para. 6, concerning affected environment. The environmental impact statement will succinctly describe the environment to be affected by the proposed action. Data and analyses in the statement will be commensurate with the importance of the impact, with less important material summarized, consolidated, or simply referenced. Effort and attention will be concentrated on important issues; useless bulk will be eliminated.

Technical Rationale

The review conducted under this ESRP leads to preparation of a section describing the affected environment for the SEIS that provides background information to be used in evaluating radioactive waste management systems associated with continued plant operations and refurbishment activities associated with license renewal.

III. REVIEW PROCEDURES

The material to be prepared on the radioactive waste management and effluent control systems is informational in nature. No specific analysis is required. The following review steps are suggested:

1. Review the discussion of radioactive waste management and effluent control systems in the *Generic Environmental Impact Statement for License Renewal of Nuclear Power Plant* (GEIS; NUREG-1437).

2. Obtain a description of the radioactive waste management and effluent control systems for the applicant's plant. The description should include identification of release points.

3. Obtain information on changes to the radioactive waste management and effluent control systems that would affect releases and exposures from continued plant operations during the license renewal term.

4. Obtain information on planned changes to the radioactive waste management and effluent control systems during refurbishment.

5. Prepare a section describing the radioactive waste management and effluent control systems for the SEIS. This section should include general descriptions of gaseous, liquid, and solid waste processing systems. It should also generally describe the applicant's gaseous and liquid effluent monitoring systems.

IV. EVALUATION FINDINGS

The depth and extent of the input to the SEIS would depend on plant- and site-specific factors. The level of detail of information included in the SEIS should be scaled according to the anticipated magnitudes of the expected impacts. The reviewer should verify that the radioactive waste management and effluent control system descriptions are consistent, accurate, and given in sufficient detail to serve the needs of the reviewers for ESRPs in other chapters.

V. IMPLEMENTATION

The method described in this ESRP would be used by the staff in evaluating conformance with the Commission's regulations, except in those cases in which the applicant for license renewal proposes an acceptable alternative for complying with specified portions of the regulations.

VI. BIBLIOGRAPHY

10 CFR Part 51. *Code of Federal Regulations*, Title 10, *Energy,* Part 51, "Environmental Protection Regulations for Domestic Licensing and Related Regulatory Functions."

U.S. Nuclear Regulatory Commission (NRC). 2013. *Generic Environmental Impact Statement for License Renewal of Nuclear Plants.* NUREG-1437, Vols. 1, 2, and 3, Revision 1. Office of Nuclear Reactor Regulation, Washington, D.C.

NUREG-1555, Supplement 1

U.S. NUCLEAR REGULATORY COMMISSION

ENVIRONMENTAL STANDARD REVIEW PLAN

OFFICE OF NUCLEAR REACTOR REGULATION

3.10.2 NONRADIOACTIVE WASTE SYSTEMS

I. AREAS OF REVIEW

This environmental standard review plan (ESRP) provides guidance for the preparation of a supplemental environmental impact statement (SEIS) section describing the applicant's nonradioactive waste management and control systems.

The scope includes describing the existing systems, describing any changes to the systems to be made during the license renewal term or refurbishment.

Data and Information Needs

The types of data and information needed would be affected by nuclear power plant site- and plant-specific factors; the level of detail should be scaled according to the anticipated magnitude of the potential impacts. The following data or information may be needed:

- a description of the nonradioactive effluent treatment systems

- identification of sources and types of nonradioactive liquid and solid waste material within the plant

- identification of principal release points for nonradioactive materials to the environment and historical information on composition of discharges

3.10.2-1

USNRC ENVIRONMENTAL STANDARD REVIEW PLAN

Environmental standard review plans are prepared for the guidance of the Office of Nuclear Reactor Regulation staff responsible for environmental reviews for nuclear power plants. These documents are made available to the public as part of the Commission's policy to inform the nuclear industry and the general public of regulatory procedures and policies. Environmental standard review plans are not substitutes for regulatory guides or the Commission's regulations and compliance with them is not required. These supplemental environmental standard review plans are keyed to Regulatory Guide 4.2, Supplement 1, "Preparation of Environmental Reports for Nuclear Power Plant License Renewal Applications."

Published environmental standard review plans will be revised periodically, as appropriate, to accommodate comments and to reflect new information and experience.

Comments and suggestions for improvement will be considered and should be sent to the U.S. Nuclear Regulatory Commission, Office of Nuclear Reactor Regulation, Washington, DC 20555-0001.

- documentation of the permits issued by the agencies responsible for permitting nonradioactive waste systems for atmospheric, liquid, or solid effluents (e.g., National Pollutant Discharge Elimination System permits)

- a description of a pollution prevention and waste minimization program, if available

- information on the changes in nonradiological impacts from operation that are expected during the renewal term

II. ACCEPTANCE CRITERIA

The reviewer should ensure that the introductory and descriptive paragraphs prepared under this ESRP are consistent with the following regulation:

- 10 CFR 51.45(d), "Status of Compliance." The environmental report shall list all Federal permits, licenses, approvals and other entitlements which must be obtained in connection with the proposed action and shall describe the status of compliance with these requirements. The environmental report shall also include a discussion of the status of compliance with applicable environmental quality standards and requirements including, but not limited to, applicable zoning and land-use regulations, and thermal and other water pollution limitations or requirements which have been imposed by Federal, State, regional, and local agencies having responsibility for environmental protection.

- 10 CFR 51.53(c)(2). The report must contain a description of the proposed action, including the applicant's plans to modify the facility or its administrative control procedures as described in accordance with 10 CFR 54.21 of this chapter. This report must describe in detail the affected environment around the plant, the modifications directly affecting the environment or any plant effluents, and any planned refurbishment activities. In addition, the applicant shall discuss in this report the environmental impacts of alternatives and any other matters discussed in 10 CFR51.45.

- 10 CFR 51.70(b). The draft environmental impact statement will be concise, clear, and analytic, and written in plain language with appropriate graphics.... The format provided in Section 1(a) of Appendix A of this subpart should be used. The Nuclear Regulatory Commission (NRC) staff will independently evaluate and be responsible for the reliability of all information used in the draft environmental impact statement.

- 10 CFR Part 51, Appendix A to Subpart A, para. 6, concerning affected environment. The environmental impact statement will succinctly describe the environment to be affected by the proposed action. Data and analyses in the statement will be commensurate with the importance of the impact, with less important material summarized, consolidated, or simply referenced. Effort and attention will be concentrated on important issues; useless bulk will be eliminated.

<u>Technical Rationale</u>

The review conducted under this ESRP leads to preparation of a section describing the affected environment for the SEIS that provides background information to be used in evaluating nonradioactive waste management systems associated with continued plant operations and refurbishment activities associated with license renewal.

III. REVIEW PROCEDURES

The material to be prepared on the nonradioactive waste management and effluent control systems is informational in nature. No specific analysis is required. The following review steps are suggested:

1. Review the discussion of nonradioactive waste systems in the *Generic Environmental Impact Statement for License Renewal of Nuclear Power Plant* (GEIS; NUREG-1437).

2. Obtain a description of the nonradioactive wastes and effluent control systems for the applicant's plant. The description should include identification of the type of waste generated, regulatory permits, release points, storage, and disposal.

3. Obtain information on changes to the nonradioactive waste and effluent control systems that would affect releases from continued plant operations during the renewal term.

4. Obtain information on planned changes to the nonradioactive waste and effluent control systems during refurbishment.

5. Obtain a description of the pollution prevention and waste minimization program or policy, if available.

Prepare a section describing the nonradioactive waste and effluent control systems for the SEIS.

IV. EVALUATION FINDINGS

The depth and extent of the input to the SEIS would depend on plant- and site-specific factors. The level of detail of information included in the SEIS should be scaled according to the anticipated magnitudes of the expected impacts. The reviewer should verify that the nonradioactive waste management and effluent control system descriptions are consistent, accurate, and given in sufficient detail to serve the needs of the reviewers for ESRPs in other chapters.

V. IMPLEMENTATION

The method described in this ESRP would be used by the staff in evaluating conformance with the Commission's regulations, except in those cases in which the applicant for license renewal proposes an acceptable alternative for complying with specified portions of the regulations.

VI. BIBLIOGRAPHY

10 CFR Part 51. *Code of Federal Regulations*, Title 10, *Energy,* Part 51, "Environmental Protection Regulations for Domestic Licensing and Related Regulatory Functions."

U.S. Nuclear Regulatory Commission (NRC). 2013. *Generic Environmental Impact Statement for License Renewal of Nuclear Plants.* NUREG-1437, Vols. 1, 2, and 3, Revision 1. Office of Nuclear Reactor Regulation, Washington, D.C.

3.11 REFERENCES

I. AREAS OF REVIEW

This environmental standard review plan (ESRP) provides guidance for the listing of references cited in the supplemental environmental impact statement (SEIS) chapter on the affected environment.

II. ACCEPTANCE CRITERIA

Acceptance criteria for the preparation of the reference list are based on the following regulation:

- 10 CFR 51.70(b). The draft environmental impact statement will be concise, clear, and analytic, will be written in plain language with appropriate graphics….The format provided in Section 1(a) of Appendix A of this subpart should be used. The Nuclear Regulatory Commission (NRC) staff will independently evaluate and be responsible for the reliability of all information used in the draft environmental impact statement.

III. REVIEW PROCEDURES

The environmental project manager (EPM) should contact reviewers for ESRPs 3.1 through 3.10 and compile a list of references cited in the SEIS sections. The citations should be checked for completeness.

IV. EVALUATION FINDINGS

The material to be prepared is informational in nature; no specific analysis of the data is required.

V. IMPLEMENTATION

The method described in this ESRP would be used by the staff in evaluating conformance with the Commission's regulations, except in those cases in which the applicant for license renewal proposes an acceptable alternative for complying with specified portions of the regulations.

VI. BIBLIOGRAPHY

10 CFR Part 51. *Code of Federal Regulations*, Title 10, *Energy,* Part 51, "Environmental Protection Regulations for Domestic Licensing and Related Regulatory Functions."

U.S. NUCLEAR REGULATORY COMMISSION
ENVIRONMENTAL STANDARD
REVIEW PLAN
OFFICE OF NUCLEAR REACTOR REGULATION

4.0 ENVIRONMENTAL CONSEQUENCES AND MITIGATING ACTIONS

I. AREAS OF REVIEW

The following sections address the environmental consequences of (1) the proposed action, which includes the potential impacts from continued plant operations and refurbishment, and (2) the environmental consequences of alternatives to the proposed action, which includes the potential impacts from the construction and operation of fossil fuel alternatives, new nuclear alternative, renewable energy alternatives, and alternatives to reduce or avoid adverse environmental impacts (e.g., constructing and operating a new cooling system). This environmental standard review plan (ESRP) provides guidance for the staff's introduction to the section of the supplemental environmental impact statement (SEIS) that reviews the environmental impacts of continued operations during the renewal term and refurbishment.

The scope of this plan introduces the material from the reviews conducted under ESRPs 4.1 through 4.12. It includes a description of the environmental issues associated with continued operation during the renewal term and refurbishment discussed in the *Generic Environmental Impact Statement for License Renewal of Nuclear Plants* (GEIS; NUREG-1437, Volumes 1, 2, and 3, Revision 1), identifies those issues that the staff has determined to be inapplicable to the applicant's plant because of plant design, and directs readers to SEIS sections that discuss the applicable issues.

II. ACCEPTANCE CRITERIA

The reviewer should ensure that the introductory paragraphs prepared under this ESRP are consistent with the following regulations:

- 10 CFR 51.45(c), "Analysis." The environmental report must include an analysis that considers and balances the environmental effects of the proposed action, the environmental impacts of replacement power alternatives, and alternatives available for reducing or avoiding adverse environmental effects.

- 10 CFR 51.53(c)(2). The report must contain a description of the proposed action, including the applicant's plans to modify the facility or its administrative control procedures as described in accordance with 10 CFR 54.21 of this chapter. This report must describe in detail the affected environment around the plant, the modifications directly affecting the environment or any plant effluents, and any planned refurbishment activities. In addition, the applicant shall discuss in this report the environmental impacts of alternatives and any other matters discussed in 10 CFR 51.45.

- 10 CFR 51.70(b). The draft environmental impact statement will be concise, clear, and analytic, and written in plain language with appropriate graphics….The format provided in Section 1(a) of Appendix A of this subpart should be used. The NRC staff will independently evaluate and be responsible for the reliability of all information used in the draft environmental impact statement.

- 10 CFR 51.71(d) concerning the draft environmental impact statement will include a preliminary analysis that considers and weighs the environmental effects of the proposed action; the environmental impacts of replacement power alternatives; and alternatives available for reducing or avoiding adverse environmental effects, among other things.

- 10 CFR 51.95(c), concerning renewal of an operating license or combined license for a nuclear power plant. Under Parts 52 or 54 of this chapter, the Commission shall prepare an environmental impact statement, which is a supplement to the Commission's NUREG-1437, "Generic Environmental Impact Statement for License Renewal of Nuclear Plants."

- 10 CFR Part 51, Appendix A to Subpart A, para. 7, concerning the environmental consequences of alternatives, including the proposed actions and any mitigating actions which may be taken. Alternatives eliminated from detailed study will be identified and a discussion of those alternatives will be confined to a brief statement of the reasons why the alternatives were eliminated. The level of information for each alternative considered in detail will reflect the depth of analysis required for sound decisionmaking.

- 10 CFR Part 51, Appendix B to Subpart A, "Environmental Effect of Renewing the Operating License of a Nuclear Power Plant," Table B-1, "Summary of Findings on NEPA Issues for License Renewal of Nuclear Power Plants."

Technical Rationale

The review conducted under this ESRP leads to the preparation of SEIS sections that incorporate the conclusions in the GEIS related to the environmental impacts of continued plant operations during the license renewal term, refurbishment, and replacement power alternatives. The review should also address any new and significant information.

III. REVIEW PROCEDURES

The material to be prepared is informational in nature; no specific analysis of data is required. Environmental issues associated with continued operations during the renewal term and refurbishment considered in the GEIS that were determined to be Category 1 or uncategorized (see General Introduction to this volume) issues are listed in Table 4.0-1.

Table 4.0-1. Category 1 and Uncategorized Issues

Table B-1. Summary of Findings on NEPA Issues for License Renewal of Nuclear Power Plants		
Issue	**Category**	**Finding**
Land Use		
Onsite land use	1	SMALL. Changes in onsite land use from continued operations and refurbishment associated with license renewal would be a small fraction of the nuclear power plant site and would involve only land that is controlled by the licensee.
Offsite land use	1	SMALL. Offsite land use would not be affected by continued operations and refurbishment associated with license renewal.
Offsite land use in transmission line right-of-ways (ROWs)	1	SMALL. Use of transmission line ROWs from continued operations and refurbishment associated with license renewal would continue with no change in land use restrictions.
Visual Resources		
Aesthetic impacts	1	SMALL. No important changes to the visual appearance of plant structures or transmission lines are expected from continued operations and refurbishment associated with license renewal.
Air Quality		
Air quality impacts (all plants)	1	SMALL. Air quality impacts from continued operations and refurbishment associated with license renewal are expected to be small at all plants. Emissions resulting from refurbishment activities at locations in or near air quality nonattainment or maintenance areas would be short-lived and would cease after these refurbishment activities are completed. Operating experience has shown that the scale of refurbishment activities has not resulted in exceedance of the de minimis thresholds for criteria pollutants, and best management practices including fugitive dust controls and the imposition of permit conditions in State and local air emissions permits would ensure conformance with applicable State or Tribal implementation plans.
		Emissions from emergency diesel generators and fire pumps and routine operations of boilers used for space heating would not be a concern, even for plants located in or adjacent to nonattainment areas. Impacts from cooling tower particulate emissions even under the worst-case situations have been small.
Air quality effects of transmission lines	1	SMALL. Production of ozone and oxides of nitrogen is insignificant and does not contribute measurably to ambient levels of these gases.
Noise		
Noise impacts	1	SMALL. Noise levels would remain below regulatory guidelines for offsite receptors during continued operations and refurbishment associated with license renewal.

Table 4.0-1. Category 1 and Uncategorized Issues (Cont.)

Table B-1. Summary of Findings on NEPA Issues for License Renewal of Nuclear Power Plants		
Issue	**Category**	**Finding**
Geologic Environment		
Geology and soils	1	SMALL. The effect of geologic and soil conditions on plant operations and the impact of continued operations and refurbishment activities on geology and soils would be small for all nuclear power plants and would not change appreciably during the license renewal term.
Surface Water Resources		
Surface-water use and quality (non-cooling-system impacts)	1	SMALL. Impacts are expected to be small if best management practices are employed to control soil erosion and spills. Surface water use associated with continued operations and refurbishment associated with license renewal would not increase significantly or would be reduced if refurbishment occurs during a plant outage.
Altered current patterns at intake and discharge structures	1	SMALL. Altered current patterns would be limited to the area in the vicinity of the intake and discharge structures. These impacts have been small at operating nuclear power plants.
Altered salinity gradients	1	SMALL. Effects on salinity gradients would be limited to the area in the vicinity of the intake and discharge structures. These impacts have been small at operating nuclear power plants.
Altered thermal stratification of lakes	1	SMALL. Effects on thermal stratification would be limited to the area in the vicinity of the intake and discharge structures. These impacts have been small at operating nuclear power plants.
Scouring caused by discharged cooling water	1	SMALL. Scouring effects would be limited to the area in the vicinity of the intake and discharge structures. These impacts have been small at operating nuclear power plants.
Discharge of metals in cooling system effluent	1	SMALL. Discharges of metals have not been found to be a problem at operating nuclear power plants with cooling-tower-based heat dissipation systems and have been satisfactorily mitigated at other plants. Discharges are monitored and controlled as part of the National Pollutant Discharge Elimination System (NPDES) permit process.
Discharge of biocides, sanitary wastes, and minor chemical spills	1	SMALL. The effects of these discharges are regulated by State and Federal environmental agencies. Discharges are monitored and controlled as part of the NPDES permit process. These impacts have been small at operating nuclear power plants.
Surface water use conflicts (plants with once-through cooling systems)	1	SMALL. These conflicts have not been found to be a problem at operating nuclear power plants with once-through heat dissipation systems.
Effects of dredging on surface water quality	1	SMALL. Dredging to remove accumulated sediments in the vicinity of intake and discharge structures and to maintain barge shipping has not been found to be a problem for surface water quality. Dredging is performed under permit from the U.S. Army Corps of Engineers, and possibly, from other State or local agencies.
Temperature effects on sediment transport capacity	1	SMALL. These effects have not been found to be a problem at operating nuclear power plants and are not expected to be a problem.

Table 4.0-1. Category 1 and Uncategorized Issues (Cont.)

Table B-1. Summary of Findings on NEPA Issues for License Renewal of Nuclear Power Plants		
Issue	**Category**	**Finding**
Groundwater Resources		
Groundwater contamination and use (non-cooling-system impacts)	1	SMALL. Extensive dewatering is not anticipated from continued operations and refurbishment associated with license renewal. Industrial practices involving the use of solvents, hydrocarbons, heavy metals, or other chemicals, and/or the use of wastewater ponds or lagoons have the potential to contaminate site groundwater, soil, and subsoil. Contamination is subject to State- or Environmental Protection Agency-regulated cleanup and monitoring programs. The application of best management practices for handling any materials produced or used during these activities would reduce impacts.
Groundwater use conflicts (plants that withdraw less than 100 gallons per minute [gpm])	1	SMALL. Plants that withdraw less than 100 gpm are not expected to cause any groundwater use conflicts.
Groundwater quality degradation resulting from water withdrawals	1	SMALL. Groundwater withdrawals at operating nuclear power plants would not contribute significantly to groundwater quality degradation.
Groundwater quality degradation (plants with cooling ponds in salt marshes)	1	SMALL. Sites with closed-cycle cooling ponds could degrade groundwater quality. However, groundwater in salt marshes is naturally brackish, and thus not potable. Consequently, the human use of such groundwater is limited to industrial purposes.
Terrestrial Resources		
Exposure of terrestrial organisms to radionuclides	1	SMALL. Doses to terrestrial organisms from continued operations and refurbishment associated with license renewal are expected to be well below exposure guidelines developed to protect these organisms.
Cooling system impacts on terrestrial resources (plants with once-through cooling systems or cooling ponds)	1	SMALL. No adverse effects to terrestrial plants or animals have been reported as a result of increased water temperatures, fogging, humidity, or reduced habitat quality. Due to the low concentrations of contaminants in cooling system effluents, uptake and accumulation of contaminants in the tissues of wildlife exposed to the contaminated water or aquatic food sources are not expected to be significant issues.
Cooling tower impacts on vegetation (plants with cooling towers)	1	SMALL. Impacts from salt drift, icing, fogging, or increased humidity associated with cooling tower operation have the potential to affect adjacent vegetation, but these impacts have been small at operating nuclear power plants and are not expected to change over the license renewal term.
Bird collisions with plant structures and transmission lines	1	SMALL. Bird collisions with cooling towers and other plant structures and transmission lines occur at rates that are unlikely to affect local or migratory populations and the rates are not expected to change.
Transmission line right-of-way (ROW) management impacts on terrestrial resources	1	SMALL. Continued ROW management during the license renewal term is expected to keep terrestrial communities in their current condition. Application of best management practices would reduce the potential for impacts.

Table 4.0-1. Category 1 and Uncategorized Issues (Cont.)

Table B-1. Summary of Findings on NEPA Issues for License Renewal of Nuclear Power Plants		
Issue	**Category**	**Finding**
Terrestrial Resources (Continued)		
Electromagnetic fields on flora and fauna (plants, agricultural crops, honeybees, wildlife, livestock)	1	SMALL. No significant impacts of electromagnetic fields on terrestrial flora and fauna have been identified. Such effects are not expected to be a problem during the license renewal term.
Aquatic Resources		
Impingement and entrainment of aquatic organisms (plants with cooling towers)	1	SMALL. Impingement and entrainment rates are lower at plants that use closed-cycle cooling with cooling towers because the rates and volumes of water withdrawal needed for makeup are minimized.
Entrainment of phytoplankton and zooplankton (all plants)	1	SMALL. Entrainment of phytoplankton and zooplankton has not been found to be a problem at operating nuclear power plants and is not expected to be a problem during the license renewal term.
Thermal impacts on aquatic organisms (plants with cooling towers)	1	SMALL. Thermal effects associated with plants that use cooling towers are expected to be small because of the reduced amount of heated discharge.
Infrequently reported thermal impacts (all plants)	1	SMALL. Continued operations during the license renewal term are expected to have small thermal impacts with respect to the following: - Cold shock has been satisfactorily mitigated at operating nuclear plants with once-through cooling systems, has not endangered fish populations or been found to be a problem at operating nuclear power plants with cooling towers or cooling ponds, and is not expected to be a problem. - Thermal plumes have not been found to be a problem at operating nuclear power plants and are not expected to be a problem. - Thermal discharge may have localized effects but is not expected to affect the larger geographical distribution of aquatic organisms. - Premature emergence has been found to be a localized effect at some operating nuclear power plants but has not been a problem and is not expected to be a problem. - Stimulation of nuisance organisms has been satisfactorily mitigated at the single nuclear power plant with a once-through cooling system where previously it was a problem. It has not been found to be a problem at operating nuclear power plants with cooling towers or cooling ponds and is not expected to be a problem.
Effects of cooling water discharge on dissolved oxygen, gas supersaturation, and eutrophication	1	SMALL. Gas supersaturation was a concern at a small number of operating nuclear power plants with once-through cooling systems but has been mitigated. Low dissolved oxygen was a concern at one nuclear power plant with a once-through cooling system but has been mitigated. Eutrophication (nutrient loading) and resulting effects on chemical and biological oxygen demands have not been found to be a problem at operating nuclear power plants.

Table 4.0-1. Category 1 and Uncategorized Issues (Cont.)

Table B-1. Summary of Findings on NEPA Issues for License Renewal of Nuclear Power Plants		
Issue	**Category**	**Finding**
Aquatic Resources – Continued		
Effects of nonradiological contaminants on aquatic organisms	1	SMALL. Best management practices and discharge limitations of NPDES permits are expected to minimize the potential for impacts to aquatic resources during continued operations and refurbishment associated with license renewal. Accumulation of metal contaminants has been a concern at a few nuclear power plants, but has been satisfactorily mitigated by replacing copper alloy condenser tubes with those of another metal.
Exposure of aquatic organisms to radionuclides	1	SMALL. Doses to aquatic organisms are expected to be well below exposure guidelines developed to protect these aquatic organisms.
Effects of dredging on aquatic organisms	1	SMALL. Dredging at nuclear power plants is expected to occur infrequently, would be of relatively short duration, and would affect relatively small areas. Dredging is performed under permit from the U.S. Army Corps of Engineers, and possibly, from other State or local agencies.
Effects on aquatic resources (non-cooling-system impacts)	1	SMALL. Licensee application of appropriate mitigation measures is expected to result in no more than small changes to aquatic communities from their current condition.
Impacts of transmission line right-of-way (ROW) management on aquatic resources	1	SMALL. Licensee application of best management practices to ROW maintenance is expected to result in no more than small impacts to aquatic resources.
Losses from predation, parasitism, and disease among organisms exposed to sublethal stresses	1	SMALL. These types of losses have not been found to be a problem at operating nuclear power plants and are not expected to be a problem during the license renewal term.
Socioeconomics		
Employment and income, recreation and tourism	1	SMALL. Impacts to employment, income, recreation, and tourism from continued operations and refurbishment associated with license renewal are expected to be small.
Tax revenues	1	SMALL. Nuclear plants provide tax revenue to local jurisdictions in the form of property tax payments, payments in lieu of tax (PILOT), or tax payments on energy production. The amount of tax revenue paid during the license renewal term as a result of continued operations and refurbishment associated with license renewal is not expected to change.

Table 4.0-1. Category 1 and Uncategorized Issues (Cont.)

Table B-1. Summary of Findings on NEPA Issues for License Renewal of Nuclear Power Plants		
Issue	**Category**	**Finding**
Socioeconomics – Continued		
Community services and education	1	SMALL. Changes resulting from continued operations and refurbishment associated with license renewal to local community and educational services would be small. With little or no change in employment at the licensee's plant, value of the power plant, payments on energy production, and PILOT payments expected during the license renewal term, community and educational services would not be affected by continued power plant operations.
Population and housing	1	SMALL. Changes resulting from continued operations and refurbishment associated with license renewal to regional population and housing availability and value would be small. With little or no change in employment at the licensee's plant expected during the license renewal term, population and housing availability and values would not be affected by continued power plant operations.
Transportation	1	SMALL. Changes resulting from continued operations and refurbishment associated with license renewal to traffic volumes would be small.
Human Health		
Radiation exposures to the public	1	SMALL. Radiation doses to the public from continued operations and refurbishment associated with license renewal are expected to continue at current levels, and would be well below regulatory limits.
Radiation exposures to plant workers	1	SMALL. Occupational doses from continued operations and refurbishment associated with license renewal are expected to be within the range of doses experienced during the current license term, and would continue to be well below regulatory limits.
Human health impact from chemicals	1	SMALL. Chemical hazards to plant workers resulting from continued operations and refurbishment associated with license renewal are expected to be minimized by the licensee implementing good industrial hygiene practices as required by permits and Federal and State regulations. Chemical releases to the environment and the potential for impacts to the public are expected to be minimized by adherence to discharge limitations of NPDES and other permits.
Microbiological hazards to plant workers	1	SMALL. Occupational health impacts are expected to be controlled by continued application of accepted industrial hygiene practices to minimize worker exposures as required by permits and Federal and State regulations.
Chronic effects of electromagnetic fields (EMFs)	N/A	UNCERTAIN IMPACT. Studies of 60-Hz EMFs have not uncovered consistent evidence linking harmful effects with field exposures. EMFs are unlike other agents that have a toxic effect (e.g., toxic chemicals and ionizing radiation) in that dramatic acute effects cannot be forced and longer-term effects, if real, are subtle. Because the state of the science is currently inadequate, no generic conclusion on human health impacts is possible.

Table 4.0-1. Category 1 and Uncategorized Issues (Cont.)

Table B-1. Summary of Findings on NEPA Issues for License Renewal of Nuclear Power Plants		
Issue	**Category**	**Finding**
Human Health – Continued		
Physical occupational hazards	1	SMALL. Occupational safety and health hazards are generic to all types of electrical generating stations, including nuclear power plants, and are of small significance if the workers adhere to safety standards and use protective equipment as required by Federal and State regulations.
Postulated Accidents		
Design-basis accidents	1	SMALL. The Nuclear Regulatory Commission (NRC) staff has concluded that the environmental impacts of design-basis accidents are of small significance for all plants.
Waste Management		
Low-level waste storage and disposal	1	SMALL. The comprehensive regulatory controls that are in place and the low public doses being achieved at reactors ensure that the radiological impacts to the environment would remain small during the license renewal term.
Onsite storage of spent nuclear fuel	1	SMALL. The expected increase in the volume of spent fuel from an additional 20 years of operation can be safely accommodated onsite during the license renewal term with small environmental effects through dry or pool storage at all plants.
Offsite radiological impacts of spent nuclear fuel and high-level waste disposal	N/A	UNCERTAIN IMPACT. The generic conclusion on offsite radiological impacts of spent nuclear fuel and high-level waste is not being finalized pending the completion of a generic environmental impact statement on waste confidence.[*]
Mixed-waste storage and disposal	1	SMALL. The comprehensive regulatory controls and the facilities and procedures that are in place ensure proper handling and storage, as well as negligible doses and exposure to toxic materials for the public and the environment at all plants. License renewal would not increase the small, continuing risk to human health and the environment posed by mixed waste at all plants. The radiological and nonradiological environmental impacts of long-term disposal of mixed waste from any individual plant at licensed sites are small.
Nonradioactive waste storage and disposal	1	SMALL. No changes to systems that generate nonradioactive waste are anticipated during the license renewal term. Facilities and procedures are in place to ensure continued proper handling, storage, and disposal, as well as negligible exposure to toxic materials for the public and the environment at all plants.
Uranium Fuel Cycle		
Offsite radiological impacts – individual impacts from other than the disposal of spent fuel and high-level waste	1	SMALL. The impacts to the public from radiological exposures have been considered by the Commission in Table S-3 of this part. Based on information in the GEIS, impacts to individuals from radioactive gaseous and liquid releases, including radon-222 and technetium-99, would remain at or below the NRC's regulatory limits.

Table 4.0-1. Category 1 and Uncategorized Issues (Cont.)

Table B-1. Summary of Findings on NEPA Issues for License Renewal of Nuclear Power Plants		
Issue	**Category**	**Finding**
Uranium Fuel Cycle – Continued		
Offsite radiological impacts – collective impacts from other than the disposal of spent fuel and high-level waste	1	There are no regulatory limits applicable to collective doses to the general public from fuel-cycle facilities. The practice of estimating health effects on the basis of collective doses may not be meaningful. All fuel-cycle facilities are designed and operated to meet the applicable regulatory limits and standards. The Commission concludes that the collective impacts are acceptable. The Commission concludes that the impacts would not be sufficiently large to require the NEPA conclusion, for any plant, that the option of extended operation under 10 CFR Part 54 should be eliminated. Accordingly, while the Commission has not assigned a single level of significance for the collective impacts of the uranium fuel cycle, this issue is considered Category 1.
Nonradiological impacts of the uranium fuel cycle	1	SMALL. The nonradiological impacts of the uranium fuel cycle resulting from the renewal of an operating license for any plant would be small.
Transportation	1	SMALL. The impacts of transporting materials to and from uranium-fuel-cycle facilities on workers, the public, and the environment are expected to be small.
Termination of Nuclear Power Plant Operations and Decommissioning		
Termination of plant operations and decommissioning on all resources	1	SMALL. License renewal is expected to have a negligible effect on the impacts of terminating operations and decommissioning on all resources.

* As a result of the decision of United States Court of Appeals in *New York v. NRC*, 681 F.3d 471 (D.C. Cir. 2012), the NRC cannot rely upon its waste confidence decision and rule until it has taken those actions that will address the deficiencies identified by the D.C. Circuit. Although the waste confidence decision and rule did not assess the impacts associated with disposal of spent nuclear fuel and high-level waste in a repository, it did reflect the Commission's confidence, at the time, in the technical feasibility of a repository and when that repository could have been expected to become available. Without the analysis in the waste confidence decision and rule regarding the technical feasibility and availability of a repository, the NRC cannot assess how long the spent fuel will need to be stored onsite.

Issues and processes common to all nuclear power plants having generic (i.e., the same or similar) environmental impacts are considered Category 1 issues. In the absence of new and significant information, the conclusions in the GEIS may be adopted in the SEIS. Category 2 issues are those issues that cannot be generically dispositioned and require a plant-specific analysis to determine the level of impact. These issues are listed in Table 4.0-2.

Table 4.0-2. Category 2 Issues

Table B-1. Summary of Findings on NEPA Issues for License Renewal of Nuclear Power Plants		
Issue	Category	Finding
Surface Water Resources		
Surface water use conflicts (plants with cooling ponds or cooling towers using makeup water from a river)	2	SMALL or MODERATE. Impacts could be of small or moderate significance, depending on makeup water requirements, water availability, and competing water demands.
Groundwater Resources		
Groundwater use conflicts (plants that withdraw more than 100 gallons per minute (gpm)	2	SMALL, MODERATE, or LARGE. Plants that withdraw more than 100 gpm could cause groundwater use conflicts with nearby groundwater users.
Groundwater use conflicts (plants with closed-cycle cooling systems that withdraw makeup water from a river)	2	SMALL, MODERATE, or LARGE. Water use conflicts could result from water withdrawals from rivers during low-flow conditions, which may affect aquifer recharge. The significance of impacts would depend on makeup water requirements, water availability, and competing water demands.
Groundwater quality degradation (plants with cooling ponds at inland sites)	2	SMALL, MODERATE, or LARGE. Inland sites with closed-cycle cooling ponds could degrade groundwater quality. The significance of the impact would depend on cooling pond water quality, site hydrogeologic conditions (including the interaction of surface water and groundwater), and the location, depth, and pump rate of water wells.
Radionuclides released to groundwater	2	SMALL or MODERATE. Leaks of radioactive liquids from plant components and pipes have occurred at numerous plants. Groundwater protection programs have been established at all operating nuclear power plants to minimize the potential impact from any inadvertent releases. The magnitude of impacts would depend on site-specific characteristics.
Terrestrial Resources		
Effects on terrestrial resources (non-cooling-system impacts)	2	SMALL, MODERATE, or LARGE. Impacts resulting from continued operations and refurbishment associated with license renewal may affect terrestrial communities. Application of best management practices would reduce the potential for impacts. The magnitude of impacts would depend on the nature of the activity, the status of the resources that could be affected, and the effectiveness of mitigation.
Water use conflicts with terrestrial resources (plants with cooling ponds or cooling towers using makeup water from a river)	2	SMALL or MODERATE. Impacts on terrestrial resources in riparian communities affected by water use conflicts could be of moderate significance.

Table 4.0-2. Category 2 Issues (Cont.)

Table B-1. Summary of Findings on NEPA Issues for License Renewal of Nuclear Power Plants

Issue	Category	Finding
Aquatic Resources		
Impingement and entrainment of aquatic organisms (plants with once-through cooling systems or cooling ponds)	2	SMALL, MODERATE, or LARGE. The impacts of impingement and entrainment are small at many plants but may be moderate or even large at a few plants with once-through and cooling-pond cooling systems, depending on cooling system withdrawal rates and volumes and the aquatic resources at the site.
Thermal impacts on aquatic organisms (plants with once-through cooling systems or cooling ponds)	2	SMALL, MODERATE, or LARGE. Most of the effects associated with thermal discharges are localized and are not expected to affect overall stability of populations or resources. The magnitude of impacts, however, would depend on site-specific thermal plume characteristics and the nature of aquatic resources in the area.
Water use conflicts with aquatic resources (plants with cooling ponds or cooling towers using makeup water from a river)	2	SMALL or MODERATE. Impacts on aquatic resources in stream communities affected by water use conflicts could be of moderate significance in some situations.
Special Status Species and Habitats		
Threatened, endangered, and protected species and essential fish habitat	2	The magnitude of impacts on threatened, endangered, and protected species, critical habitat, and essential fish habitat would depend on the occurrence of listed species and habitats and the effects of power plant systems on them. Consultation with appropriate agencies would be needed to determine whether special status species or habitats are present and whether they would be adversely affected by continued operations and refurbishment associated with license renewal.
Historic and Cultural Resources		
Historic and cultural resources	2	Continued operations and refurbishment associated with license renewal are expected to have no more than small impacts on historic and cultural resources located onsite and in the transmission line ROW because most impacts could be mitigated by avoiding those resources. The National Historic Preservation Act (NHPA) requires the Federal agency to consult with the State Historic Preservation Officer (SHPO) and appropriate Native American Tribes to determine the potential effects on historic properties and mitigation, if necessary.
Human Health		
Microbiological hazards to the public (plants with cooling ponds or canals or cooling towers that discharge to a river)	2	SMALL, MODERATE, or LARGE. These organisms are not expected to be a problem at most operating plants except possibly at plants using cooling ponds, lakes, or canals, or that discharge into rivers. Impacts would depend on site-specific characteristics.
Electric shock hazards	2	SMALL, MODERATE, or LARGE. Electrical shock potential is of small significance for transmission lines that are operated in adherence with the National Electrical Safety Code (NESC). Without a review of conformance with NESC criteria of each nuclear power plant's in-scope transmission lines, it is not possible to determine the significance of the electrical shock potential.

Table 4.0-2. Category 2 Issues (Cont.)

Table B-1. Summary of Findings on NEPA Issues for License Renewal of Nuclear Power Plants		
Issue	Category	Finding
Postulated Accidents		
Severe accidents	2	SMALL. The probability-weighted consequences of atmospheric releases, fallout onto open bodies of water, releases to groundwater, and societal and economic impacts from severe accidents are small for all plants. However, alternatives to mitigate severe accidents must be considered for all plants that have not considered such alternatives.
Environmental Justice		
Minority and low-income populations	2	Impacts to minority and low-income populations and subsistence consumption resulting from continued operations and refurbishment associated with license renewal will be addressed in plant-specific reviews. See NRC Policy Statement on the Treatment of Environmental Justice Matters in NRC Regulatory and Licensing Actions (69 FR 52040, August 24, 2004).
Cumulative Impacts		
Cumulative impacts	2	Cumulative impacts of continued operations and refurbishment associated with license renewal must be considered on a plant-specific basis. Impacts would depend on regional resource characteristics, the resource-specific impacts of license renewal, and the cumulative significance of other factors affecting the resource.

IV. EVALUATION FINDINGS

The environmental project manager (EPM) should prepare the introductory paragraphs for the SEIS. The paragraph(s) should introduce the issues to be covered by ESRPs 4.1 through 4.12.

V. IMPLEMENTATION

The method described in this ESRP would be used by the staff in evaluating conformance with the Commission's regulations, except in those cases in which the applicant for license renewal proposes an acceptable alternative for complying with specified portions of the regulations.

VI. BIBLIOGRAPHY

10 CFR Part 51. *Code of Federal Regulations*, Title 10, *Energy,* Part 51, "Environmental Protection Regulations for Domestic Licensing and Related Regulatory Functions."

10 CFR Part 54. *Code of Federal Regulations*, Title 10, *Energy,* Part 54, "Requirements for Renewal of Operating Licenses for Nuclear Power Plants."

69 FR 52040. U.S. Nuclear Regulatory Commission. Policy Statement on the Treatment of Environmental Justice Matters in NRC Regulatory and Licensing Actions. November 5, 2003.

U.S. Nuclear Regulatory Commission (NRC). 2013. *Generic Environmental Impact Statement for License Renewal of Nuclear Plants.* NUREG-1437, Vols. 1, 2, and 3, Revision 1. Office of Nuclear Reactor Regulation, Washington, D.C.

U.S. NUCLEAR REGULATORY COMMISSION
ENVIRONMENTAL STANDARD REVIEW PLAN
OFFICE OF NUCLEAR REACTOR REGULATION

4.1 LAND USE AND VISUAL RESOURCES

I. AREAS OF REVIEW

This environmental standard review plan (ESRP) provides guidance for the review of potential land-use and visual impacts of continued plant operations and refurbishment associated with license renewal. Impacts are discussed in the *Generic Environmental Impact Statement for License Renewal of Nuclear Plants* (GEIS; NUREG-1437, Volumes 1, 2, and 3, Revision 1).

The scope includes (1) review of the discussion of potential land-use and visual impacts of the plant during the license renewal term in the GEIS, (2) review the applicant's environmental report (ER), (3) identifying and addressing any new and significant information, and (4) preparing input to the supplemental environmental impact statement (SEIS).

<u>Data and Information Needs</u>

The types of data and information needed would be affected by nuclear power plant site- and plant-specific factors. The following data or information may be needed:

- a description of the applicant's process for identifying new and potentially significant information

- any new information included in the environmental report (ER) on the land-use and visual impacts of the plant known to the applicant

---

- new and potentially significant information on the land-use and visual impacts of the plant during the license renewal term identified by the public

II. ACCEPTANCE CRITERIA

Acceptance criteria for the evaluation of land-use impacts are based on the following regulations:

- 10 CFR 51.45(c), "Analysis." The environmental report must include an analysis that considers and balances the environmental effects of the proposed action, the environmental impacts of replacement power alternatives, and alternatives available for reducing or avoiding adverse environmental effects.

- 10 CFR 51.53(c)(2). The report must contain a description of the proposed action, including the applicant's plans to modify the facility or its administrative control procedures as described in accordance with 10 CFR 54.21 of this chapter. This report must describe in detail the affected environment around the plant, the modifications directly affecting the environment or any plant effluents, and any planned refurbishment activities. In addition, the applicant shall discuss in this report the environmental impacts of alternatives and any other matters discussed in 10 CFR 51.45.

- 10 CFR 51.70(b). The draft environmental impact statement will be concise, clear, and analytic, and written in plain language with appropriate graphics....The format provided in Section 1(a) of Appendix A of this subpart should be used. The Nuclear Regulatory Commission (NRC) staff will independently evaluate and be responsible for the reliability of all information used in the draft environmental impact statement.

- 10 CFR 51.71(d), concerning the draft environmental impact statement, will include a preliminary analysis that considers and weighs the environmental effects of the proposed action, the environmental impacts of replacement power alternatives, and alternatives available for reducing or avoiding adverse environmental effects.

- 10 CFR 51.95(c), concerning the renewal of an operating license or combined license for a nuclear power plant. Under Parts 52 or 54 of this chapter, the Commission shall prepare an environmental impact statement, which is a supplement to the Commission's NUREG-1437, "Generic Environmental Impact Statement for License Renewal of Nuclear Plants."

- 10 CFR Part 51, Appendix A to Subpart A, para. 7, concerning the environmental consequences of alternatives, including the proposed actions and any mitigating actions which may be taken. Alternatives eliminated from detailed study will be identified and a discussion of those alternatives will be confined to a brief statement of the reasons why the alternatives were eliminated. The level of information for each alternative considered in detail will reflect the depth of analysis required for sound decisionmaking.

- 10 CFR Part 51, Appendix B to Subpart A, "Environmental Effect of Renewing the Operating License of a Nuclear Power Plant," Table B-1, "Summary of Findings on NEPA Issues for License Renewal of Nuclear Power Plants"

<u>Technical Rationale</u>

The review conducted under this ESRP leads to the preparation of supplemental environmental impact statement (SEIS) sections that incorporate the conclusions related to the land-use and visual impacts of continued plant operations during the license renewal term, refurbishment, and replacement power alternatives. The review should also address any new and significant information.

III. REVIEW PROCEDURES

The following review steps are suggested:

1. Review the discussion of land-use and visual impacts in the GEIS to identify the information considered and the conclusions reached. This step establishes the basis for evaluating information identified by the applicant, the public, and the staff. The following table lists the renewal term transmission line issues considered in the GEIS.

Table B-1. Summary of Findings on NEPA Issues for License Renewal of Nuclear Power Plants		
Issue	Category	Finding
Land Use		
Onsite land use	1	SMALL. Changes in onsite land use from continued operations and refurbishment associated with license renewal would be a small fraction of the nuclear power plant site and would involve only land that is controlled by the licensee.
Offsite land use	1	SMALL. Offsite land use would not be affected by continued operations and refurbishment associated with license renewal.
Offsite land use in transmission line right-of-ways (ROWs)	1	SMALL. Use of transmission line ROWs from continued operations and refurbishment associated with license renewal would continue with no change in land use restrictions.
Visual Resources		
Aesthetic impacts	1	SMALL. No important changes to the visual appearance of plant structures or transmission lines are expected from continued operations and refurbishment associated with license renewal.

2. Determine if there is new information on these issues that should be evaluated. The following sources of information should be included in the search for new information:

    *   The applicant's ER. An applicant is required by 10 CFR 51.53(c)(3)(iv) to disclose new and significant information regarding the environmental impacts of license renewal of which it is aware. In reviewing the applicant's ER, consider the applicant's process for discovering new information and evaluating the significance of any new information discovered.

    *   Records of public scoping meetings and correspondence related to the application. Compare information presented by the public with information considered in the GEIS.

    *   Land-use requirements affecting the land use of the plant.

3. Evaluate the significance of new information.

4. Prepare a section for the SEIS describing the search for new information, summarizing new information found, presenting results of evaluation of significance, and adopting conclusions from the GEIS modified as necessary to account for significant new information.

IV. EVALUATION FINDINGS

The depth and extent of the input to the SEIS would be determined by the analysis required to reach a conclusion related to the potential land-use and visual impacts of continued plant operations during the license renewal term and refurbishment. The information that should be included in the SEIS is described in the review procedures.

V. IMPLEMENTATION

The method described in this ESRP would be used in evaluating conformance with the Commission's regulations, except in those cases in which the applicant for license renewal proposes an acceptable alternative for complying with specified portions of the regulations.

VI. BIBLIOGRAPHY

10 CFR Part 51. *Code of Federal Regulations*, Title 10, *Energy,* Part 51, "Environmental Protection Regulations for Domestic Licensing and Related Regulatory Functions."

National Environmental Policy Act of 1969 (NEPA). 42 USC 4321 et seq.

U.S. Nuclear Regulatory Commission (NRC). 2013. *Generic Environmental Impact Statement for License Renewal of Nuclear Plants.* NUREG-1437, Vols. 1, 2, and 3, Revision 1. Office of Nuclear Reactor Regulation, Washington, D.C.

U.S. NUCLEAR REGULATORY COMMISSION
ENVIRONMENTAL STANDARD REVIEW PLAN
OFFICE OF NUCLEAR REACTOR REGULATION

4.2 AIR QUALITY AND NOISE

I. AREAS OF REVIEW

This environmental standard review plan (ESRP) provides guidance for the review of air quality and noise impacts from continued plant operations during the license renewal term and refurbishment. Air quality and noise impacts are discussed in the *Generic Environmental Impact Statement for License Renewal of Nuclear Plants* (GEIS; NUREG-1437, Volumes 1, 2, and 3, Revision 1).

The scope includes (1) review of the discussion of air quality and noise impacts in the GEIS, (2) review of the applicant's environmental report (ER), (3) identifying and addressing any new and significant information, and (4) preparing input to the supplemental environmental impact statement (SEIS).

Projected air quality impacts from continued operations and refurbishment is a Category 1 issue in the GEIS and Table B-1 of Appendix B to Subpart A of Part 51. Air quality effects of transmission lines and noise impacts are also Category 1 issues.

Data and Information Needs

The types of data and information needed would be affected by nuclear power plant site- and plant-specific factors. The following data or information may be needed:

*   the applicant's ER

*   the GEIS

USNRC ENVIRONMENTAL STANDARD REVIEW PLAN

Environmental standard review plans are prepared for the guidance of the Office of Nuclear Reactor Regulation staff responsible for environmental reviews for nuclear power plants. These documents are made available to the public as part of the Commission's policy to inform the nuclear industry and the general public of regulatory procedures and policies. Environmental standard review plans are not substitutes for regulatory guides or the Commission's regulations and compliance with them is not required. These supplemental environmental standard review plans are keyed to Regulatory Guide 4.2, Supplement 1, "Preparation of Environmental Reports for Nuclear Power Plant License Renewal Applications."

Published environmental standard review plans will be revised periodically, as appropriate, to accommodate comments and to reflect new information and experience.

Comments and suggestions for improvement will be considered and should be sent to the U.S. Nuclear Regulatory Commission, Office of Nuclear Reactor Regulation, Washington, DC 20555-0001.

• new information on the air quality impacts identified by the public and other information sources

II. ACCEPTANCE CRITERIA

Acceptance criteria for the evaluation of air quality and noise impacts are based on the following regulations:

• 10 CFR 51.45(c), "Analysis." The environmental report must include an analysis that considers and balances the environmental effects of the proposed action, the environmental impacts of alternatives to the proposed action, and alternatives available for reducing or avoiding adverse environmental effects.

• 10 CFR 51.53(c)(2). The report must contain a description of the proposed action, including the applicant's plans to modify the facility or its administrative control procedures as described in accordance with 10 CFR 54.21 of this chapter. This report must describe in detail the affected environment around the plant, the modifications directly affecting the environment or any plant effluents, and any planned refurbishment activities. In addition, the applicant shall discuss in this report the environmental impacts of alternatives and any other matters discussed in 10 CFR 51.45.

• 10 CFR 51.70(b). The draft environmental impact statement will be concise, clear, and analytic, and written in plain language with appropriate graphics....The format provided in Section 1(a) of Appendix A of this subpart should be used. The NRC staff will independently evaluate and be responsible for the reliability of all information used in the draft environmental impact statement.

• 10 CFR 51.71(d), concerning the draft environmental impact statement, will include a preliminary analysis that considers and weighs the environmental effects of the proposed action, the environmental impacts of alternatives to the proposed action, and alternatives available for reducing or avoiding adverse environmental effects.

• 10 CFR 51.95(c), concerning renewal of an operating license or combined license for a nuclear power plant. Under Parts 52 or 54 of this chapter, the Commission shall prepare an environmental impact statement, which is a supplement to the Commission's NUREG-1437, "Generic Environmental Impact Statement for License Renewal of Nuclear Plants."

• 10 CFR Part 51, Appendix A to Subpart A, para. 7, concerning the environmental consequences of alternatives, including the proposed actions and any mitigating actions which may be taken. Alternatives eliminated from detailed study will be identified and a discussion of those alternatives will be confined to a brief statement of the reasons why the alternatives were eliminated. The level of information for each alternative considered in detail will reflect the depth of analysis required for sound decisionmaking.

• 10 CFR Part 51, Appendix B to Subpart A, "Environmental Effect of Renewing the Operating License of a Nuclear Power Plant," Table B-1, "Summary of Findings on NEPA Issues for License Renewal of Nuclear Power Plants"

Technical Rationale

The review conducted under this ESRP leads to the preparation of SEIS sections that incorporate the conclusions in the GEIS related to air quality and noise impacts of continued plant operations during the license renewal term, refurbishment, and alternatives to the proposed action. The review should also address any new and significant information.

Table B-1. Summary of Findings on NEPA Issues for License Renewal of Nuclear Power Plants		
Issue	Category	Finding
Air Quality		
Air quality impacts (all plants)	1	SMALL. Air quality impacts from continued operations and refurbishment associated with license renewal are expected to be small at all plants. Emissions resulting from refurbishment activities at locations in or near air quality nonattainment or maintenance areas would be short-lived and would cease after these refurbishment activities are completed. Operating experience has shown that the scale of refurbishment activities has not resulted in exceedance of the de minimis thresholds for criteria pollutants, and best management practices including fugitive dust controls and the imposition of permit conditions in State and local air emissions permits would ensure conformance with applicable State or Tribal implementation plans.
		Emissions from emergency diesel generators and fire pumps and routine operations of boilers used for space heating would not be a concern, even for plants located in or adjacent to nonattainment areas. Impacts from cooling tower particulate emissions even under the worst-case situations have been small.
Air quality effects of transmission lines	1	SMALL. Production of ozone and oxides of nitrogen is insignificant and does not contribute measurably to ambient levels of these gases.
Noise		
Noise impacts	1	SMALL. Noise levels would remain below regulatory guidelines for offsite receptors during continued operations and refurbishment associated with license renewal.

III. REVIEW PROCEDURES

Suggested steps for the review process are as follows:

1. Review the discussion of air quality and noise impacts in the GEIS to identify the information considered and the conclusions reached. This step establishes the basis for evaluating information identified by the applicant, the public, and the staff. The following table lists the air quality and noise issues addressed in the GEIS.

2. Determine if there is new information on these issues that should be evaluated. The following sources of information should be included in the search for new information:

   • The applicant's ER. An applicant is required by 10 CFR 51.53(c)(3)(iv) to disclose new and significant information regarding the environmental impacts of license renewal of which it is aware. In reviewing the applicant's ER, consider the applicant's process for discovering new information and evaluating the significance of any new information discovered.

- Records of public scoping meetings and correspondence related to the application. Compare information presented by the public with information considered in the GEIS.

3. Evaluate the significance of new information.

4. Prepare a section for the SEIS describing the search for new information, summarizing new information found, presenting results of evaluation of significance, and adopting conclusions from the GEIS modified as necessary to account for new and significant information.

IV. EVALUATION FINDINGS

The depth and extent of the input to the SEIS would be determined by the analysis required to reach a conclusion related to the potential air quality impacts, effects of in-scope transmission lines, and noise impacts from continued plant operations and refurbishment. The information that should be included in the SEIS is described in the review procedures.

V. IMPLEMENTATION

The method described in this ESRP would be used by the staff in evaluating conformance with the Commission's regulations, except in those cases in which the applicant for license renewal proposes an acceptable alternative for complying with specified portions of the regulations.

VI. BIBLIOGRAPHY

10 CFR Part 51. *Code of Federal Regulations*, Title 10, *Energy,* Part 51, "Environmental Protection Regulations for Domestic Licensing and Related Regulatory Functions."

National Environmental Policy Act of 1969 (NEPA). 42 USC 4321 et seq.

U.S. Nuclear Regulatory Commission (NRC). 2013. *Generic Environmental Impact Statement for License Renewal of Nuclear Plants.* NUREG-1437, Vols. 1, 2, and 3, Revision 1. Office of Nuclear Reactor Regulation, Washington, D.C.

4.3 GEOLOGY AND SOILS

I. AREAS OF REVIEW

This environmental standard review plan (ESRP) provides guidance for the review of potential impacts of continued plant operations during the license renewal term and refurbishment associated with geology and soils. Impacts are discussed in the *Generic Environmental Impact Statement for License Renewal of Nuclear Plants* (GEIS; NUREG-1437, Volumes 1, 2, and 3, Revision 1).

The scope includes (1) review of the discussion of geology and soils in the GEIS, (2) review of the applicant's environmental report (ER), (3) identifying and addressing any new and significant information, and (4) preparing input to the supplemental environmental impact statement (SEIS).

Data and Information Needs

The types of data and information needed would be affected by nuclear power plant site- and plant-specific factors. The following data or information may be needed:

- the applicant's ER

- the GEIS

- new information on geology and soils identified by the public and other information sources

---

USNRC ENVIRONMENTAL STANDARD REVIEW PLAN

Environmental standard review plans are prepared for the guidance of the Office of Nuclear Reactor Regulation staff responsible for environmental reviews for nuclear power plants. These documents are made available to the public as part of the Commission's policy to inform the nuclear industry and the general public of regulatory procedures and policies. Environmental standard review plans are not substitutes for regulatory guides or the Commission's regulations and compliance with them is not required. These supplemental environmental standard review plans are keyed to Regulatory Guide 4.2, Supplement 1, "Preparation of Environmental Reports for Nuclear Power Plant License Renewal Applications."

Published environmental standard review plans will be revised periodically, as appropriate, to accommodate comments and to reflect new information and experience.

Comments and suggestions for improvement will be considered and should be sent to the U.S. Nuclear Regulatory Commission, Office of Nuclear Reactor Regulation, Washington, DC 20555-0001.

II. ACCEPTANCE CRITERIA

Acceptance criteria for the evaluation of geology and soil impacts are based on the following regulations:

- 10 CFR 51.45(c), "Analysis." The environmental report must include an analysis that considers and balances the environmental effects of the proposed action, the environmental impacts of alternatives to the proposed action, and alternatives available for reducing or avoiding adverse environmental effects.

- 10 CFR 51.53(c)(2). The report must contain a description of the proposed action, including the applicant's plans to modify the facility or its administrative control procedures as described in accordance with 10 CFR 54.21 of this chapter. This report must describe in detail the affected environment around the plant, the modifications directly affecting the environment or any plant effluents, and any planned refurbishment activities. In addition, the applicant shall discuss in this report the environmental impacts of alternatives and any other matters discussed in 10 CFR 51.45.

- 10 CFR 51.70(b). The draft environmental impact statement, will be concise, clear, and analytic, and written in plain language with appropriate graphics....The format provided in Section 1(a) of Appendix A of this subpart should be used. The NRC staff will independently evaluate and be responsible for the reliability of all information used in the draft environmental impact statement.

- 10 CFR 51.71(d), concerning the draft environmental impact statement, will include a preliminary analysis that considers and weighs the environmental effects of the proposed action, the environmental impacts of alternatives to the proposed action, and alternatives available for reducing or avoiding adverse environmental effects.

- 10 CFR 51.95(c), concerning the renewal of an operating license or combined license for a nuclear power plant. Under Parts 52 or 54 of this chapter, the Commission shall prepare an environmental impact statement, which is a supplement to the Commission's NUREG-1437, "Generic Environmental Impact Statement for License Renewal of Nuclear Plants."

- 10 CFR Part 51, Appendix A to Subpart A, para. 7, concerning the environmental consequences of alternatives, including the proposed actions and any mitigating actions which may be taken. Alternatives eliminated from detailed study will be identified and a discussion of those alternatives will be confined to a brief statement of the reasons why the alternatives were eliminated. The level of information for each alternative considered in detail will reflect the depth of analysis required for sound decisionmaking.

- 10 CFR Part 51, Appendix B to Subpart A, "Environmental Effect of Renewing the Operating License of a Nuclear Power Plant," Table B-1, "Summary of Findings on NEPA Issues for License Renewal of Nuclear Power Plants"

<u>Technical Rationale</u>

The review conducted under this ESRP leads to the preparation of SEIS sections that incorporate the conclusions in the GEIS related to geology and soils impacts of continued plant operations during the license renewal term, refurbishment, and alternatives to the proposed action. The review should also address any new and significant information.

III. REVIEW PROCEDURES

Suggested steps for the review process are as follows:

1. Review the discussion of geology and soils impacts in the GEIS to identify the information considered and the conclusions reached. This step establishes the basis for evaluating information identified by the applicant, the public, and the staff. The following table lists the geology and soils issue addressed in the GEIS.

Table B-1. Summary of Findings on NEPA Issues for License Renewal of Nuclear Power Plants		
Issue	**Category**	**Finding**
Geologic Environment		
Geology and soils	1	SMALL. The effect of geologic and soil conditions on plant operations and the impact of continued operations and refurbishment activities on geology and soils would be small for all nuclear power plants and would not change appreciably during the license renewal term.

2. Determine if there is new information on these issues that should be evaluated. The following sources of information should be included in the search for new information:

    • The applicant's ER. An applicant is required by 10 CFR 51.53(c)(3)(iv) to disclose new and significant information regarding the environmental impacts of license renewal of which it is aware. In reviewing the applicant's ER, consider the applicant's process for discovering new information and evaluating the significance of any new information discovered.

    • Records of public scoping meetings and correspondence related to the application. Compare information presented by the public with information considered in the GEIS.

3. Evaluate the significance of new information.

4. Prepare a section for the SEIS describing the search for new information, summarizing new information found, presenting results of evaluation of significance, and adopting conclusions from the GEIS modified as necessary to account for new and significant information.

IV. EVALUATION FINDINGS

The depth and extent of the input to the SEIS would be determined by the analysis required to reach a conclusion related to the potential geology and soils impacts from continued plant operations during the license renewal term and refurbishment. The information that should be included in the SEIS is described in the review procedures.

V. IMPLEMENTATION

The method described in this ESRP would be used in evaluating conformance with the Commission's regulations, except in those cases in which the applicant for license renewal proposes an acceptable alternative for complying with specified portions of the regulations.

VI. BIBLIOGRAPHY

10 CFR Part 51. *Code of Federal Regulations*, Title 10, *Energy*, Part 51, "Environmental Protection Regulations for Domestic Licensing and Related Regulatory Functions."

National Environmental Policy Act of 1969 (NEPA). 42 USC 4321 et seq.

U.S. Nuclear Regulatory Commission (NRC). 2013. *Generic Environmental Impact Statement for License Renewal of Nuclear Plants*. NUREG-1437, Vols. 1, 2, and 3, Revision 1. Office of Nuclear Reactor Regulation, Washington, D.C.

U.S. NUCLEAR REGULATORY COMMISSION
ENVIRONMENTAL STANDARD REVIEW PLAN
OFFICE OF NUCLEAR REACTOR REGULATION

4.4 WATER RESOURCES

I. AREAS OF REVIEW

This environmental standard review plan (ESRP) provides guidance for the analysis of surface water and groundwater impacts from continued plant operations during the license renewal term and refurbishment. Surface water and groundwater impacts are discussed in the *Generic Environmental Impact Statement for License Renewal of Nuclear Plants* (GEIS; NUREG-1437, Volumes 1, 2, and 3, Revision 1).

The scope includes (1) review the discussion of surface water and groundwater issues in the GEIS, (2) review the applicant's environmental report (ER), (3) identify and address any new and significant information, and (4) prepare input to the supplemental environmental impact statement (SEIS).

II. ACCEPTANCE CRITERIA (General for all water resource issues)

Acceptance criteria for the evaluation of surface water and groundwater impacts are based on the following regulations:

* 10 CFR 51.45(c), "Analysis." The environmental report must include an analysis that considers and balances the environmental effects of the proposed action, the environmental impacts of alternatives to the proposed action, and alternatives available for reducing or avoiding adverse environmental effects.

USNRC ENVIRONMENTAL STANDARD REVIEW PLAN

- 10 CFR 51.53(c)(2). The report must contain a description of the proposed action, including the applicant's plans to modify the facility or its administrative control procedures as described in accordance with 10 CFR 54.21 of this chapter. This report must describe in detail the affected environment around the plant, the modifications directly affecting the environment or any plant effluents, and any planned refurbishment activities. In addition, the applicant shall discuss in this report the environmental impacts of alternatives and any other matters discussed in 10 CFR 51.45.

- 10 CFR 51.53(c)(3)(ii)(A). If the applicant's plant utilizes cooling towers or cooling ponds and withdraws makeup water from a river, an assessment of the impact of the proposed action on water availability and competing water demands, the flow of the river, and related impacts on stream (aquatic) and riparian (terrestrial) ecological communities must be provided. The applicant shall also provide an assessment of the impacts of the withdrawal of water from the river on alluvial aquifers during low flow.

- 10 CFR 51.53(c)(3)(ii)(C). If the applicant's plant pumps more than 100 gallons (total onsite) of groundwater per minute, an assessment of the impact of the proposed action on groundwater must be provided.

- 10 CFR 51.53(c)(3)(ii)(D). If the applicant's plant is located at an inland site and utilizes cooling ponds, an assessment of the impact of the proposed action on groundwater quality must be provided.

- 10 CFR 51.53(c)(3)(ii)(P). An applicant shall assess the impact of any documented inadvertent releases of radionuclides into groundwater. The applicant shall include in its assessment a description of any groundwater protection program used for the surveillance of piping and components containing radioactive liquids for which a pathway to groundwater may exist. The assessment must also include a description of any past inadvertent releases and the projected impact to the environment (e.g., aquifers, rivers, lakes, ponds, the ocean) during the license renewal term.

- 10 CFR 51.70(b). The draft environmental impact statement will be concise, clear, and analytic, and written in plain language with appropriate graphics....The format provided in Section 1(a) of Appendix A of this subpart should be used. The NRC staff will independently evaluate and be responsible for the reliability of all information used in the draft environmental impact statement.

- 10 CFR 51.71(d), concerning the draft environmental impact statement, will include a preliminary analysis that considers and weighs the environmental effects of the proposed action; the environmental impacts of alternatives to the proposed action; and alternatives available for reducing or avoiding adverse environmental effects.

- 10 CFR 51.95(c), concerning renewal of an operating license or combined license for a nuclear power plant. Under Parts 52 or 54 of this chapter, the Commission shall prepare an environmental impact statement, which is a supplement to the Commission's NUREG-1437, "Generic Environmental Impact Statement for License Renewal of Nuclear Plants."

- 10 CFR Part 51, Appendix A to Subpart A, para. 7, concerning the environmental consequences of alternatives, including the proposed actions and any mitigating actions which may be taken.

Alternatives eliminated from detailed study will be identified and a discussion of those alternatives will be confined to a brief statement of the reasons why the alternatives were eliminated. The level of information for each alternative considered in detail will reflect the depth of analysis required for sound decisionmaking.

- 10 CFR Part 51, Appendix B to Subpart A, "Environmental Effect of Renewing the Operating License of a Nuclear Power Plant," Table B-1, "Summary of Findings on NEPA Issues for License Renewal of Nuclear Power Plants"

- 40 CFR Part 6, Appendix A, concerning procedures on floodplain management and wetlands protection

- Federal, State, regional, local, and affected American Indian Tribal agencies' water laws and water rights

- 40 CFR Part 122, concerning the National Pollutant Discharge Elimination System (NPDES) permit conditions for discharges including stormwater discharges

- 40 CFR Part 124, concerning the NPDES permit process

- 40 CFR Part 125, concerning water-quality standards for the NPDES

- 40 CFR Part 133, concerning treated effluents

- 40 CFR Part 149, concerning possible supplemental restrictions on waste disposal and water use in or above a sole source aquifer

- 40 CFR Part 165, concerning the disposal and storage of pesticides

- 40 CFR Part 403, concerning pretreatment of waste effluents

- 40 CFR Part 423, concerning effluent limitations on existing and new point sources

- 40 CFR Parts 700–716, concerning practices and procedures for managing toxic chemicals

Additional regulatory positions and specific criteria in support of regulations identified above are as follows:

Compliance with environmental quality standards and requirements of the Federal Water Pollution Control Act (FWPCA), commonly referred to as the Clean Water Act, is not a substitute for and does not negate the requirement for NRC to weigh the environmental impacts of the proposed action, including any degradation of water quality, and to consider alternatives to the proposed action that are available for reducing the adverse impacts. If an environmental assessment of aquatic impacts is available from the permitting authority, the NRC should consider the assessment in its determination of the magnitude of the environmental impacts in striking an overall benefit-cost balance. When no such assessment of aquatic impacts is available from the permitting authority, the

NRC (to the degree possible in conjunction with the permitting authority and other agencies having relevant expertise) should establish its own impact determination.

In *PUD No. 1 of Jefferson County v. Washington Department of Ecology*, 511 U.S. 700 (1994), the United States Supreme Court interpreted the Clean Water Act as allowing States to impose conditions on certifications, such as limitations on a given project, insofar as necessary to enforce a designated use contained in the State's water quality standard. The Court held that these limitations do not have to be specifically tied to a discharge requirement.

Technical Rationale

The review conducted under this ESRP leads to the preparation of SEIS sections that incorporate the conclusions in the GEIS related to surface water and groundwater impacts of continued plant operations during the license renewal term, refurbishment, and alternatives to the proposed action. The review should also address any new and significant information.

III. REVIEW PROCEDURES (General for all water resource issues)

Suggested steps for the review process are as follows:

1. Review the discussion of surface water and groundwater issues in the GEIS to identify the information considered and the conclusions reached. This step establishes the basis for evaluating information identified by the applicant, the public, and the staff. Category 1 surface water and groundwater issues identified in the GEIS are listed in the following table.

Table B-1. Summary of Findings on NEPA Issues for License Renewal of Nuclear Power Plants		
Issue	**Category**	**Finding**
Surface Water Resources		
Surface-water use and quality (non-cooling-system impacts)	1	SMALL. Impacts are expected to be small if best management practices are employed to control soil erosion and spills. Surface water use associated with continued operations and refurbishment associated with license renewal would not increase significantly or would be reduced if refurbishment occurs during a plant outage.
Altered current patterns at intake and discharge structures	1	SMALL. Altered current patterns would be limited to the area in the vicinity of the intake and discharge structures. These impacts have been small at operating nuclear power plants.
Altered salinity gradients	1	SMALL. Effects on salinity gradients would be limited to the area in the vicinity of the intake and discharge structures. These impacts have been small at operating nuclear power plants.
Altered thermal stratification of lakes	1	SMALL. Effects on thermal stratification would be limited to the area in the vicinity of the intake and discharge structures. These impacts have been small at operating nuclear power plants.
Scouring caused by discharged cooling water	1	SMALL. Scouring effects would be limited to the area in the vicinity of the intake and discharge structures. These

Table B-1. Summary of Findings on NEPA Issues for License Renewal of Nuclear Power Plants		
Issue	**Category**	**Finding**
		impacts have been small at operating nuclear power plants.
Discharge of metals in cooling system effluent	1	SMALL. Discharges of metals have not been found to be a problem at operating nuclear power plants with cooling-tower-based heat dissipation systems and have been satisfactorily mitigated at other plants. Discharges are monitored and controlled as part of the National Pollutant Discharge Elimination System (NPDES) permit process.
Discharge of biocides, sanitary wastes, and minor chemical spills	1	SMALL. The effects of these discharges are regulated by State and Federal environmental agencies. Discharges are monitored and controlled as part of the NPDES permit process. These impacts have been small at operating nuclear power plants.
Surface water use conflicts (plants with once-through cooling systems)	1	SMALL. These conflicts have not been found to be a problem at operating nuclear power plants with once-through heat dissipation systems.
Effects of dredging on surface water quality	1	SMALL. Dredging to remove accumulated sediments in the vicinity of intake and discharge structures and to maintain barge shipping has not been found to be a problem for surface water quality. Dredging is performed under permit from the U.S. Army Corps of Engineers, and possibly, from other State or local agencies.
Temperature effects on sediment transport capacity	1	SMALL. These effects have not been found to be a problem at operating nuclear power plants and are not expected to be a problem.
Groundwater Resources		
Groundwater contamination and use (non-cooling-system impacts)	1	SMALL. Extensive dewatering is not anticipated from continued operations and refurbishment associated with license renewal. Industrial practices involving the use of solvents, hydrocarbons, heavy metals, or other chemicals, and/or the use of wastewater ponds or lagoons have the potential to contaminate site groundwater, soil, and subsoil. Contamination is subject to State or Environmental Protection Agency regulated cleanup and monitoring programs. The application of best management practices for handling any materials produced or used during these activities would reduce impacts.
Groundwater use conflicts (plants that withdraw less than 100 gallons per minute [gpm])	1	SMALL. Plants that withdraw less than 100 gpm are not expected to cause any groundwater use conflicts.
Groundwater quality degradation resulting from water withdrawals	1	SMALL. Groundwater withdrawals at operating nuclear power plants would not contribute significantly to groundwater quality degradation.
Groundwater quality degradation (plants with cooling ponds in salt marshes)	1	SMALL. Sites with closed-cycle cooling ponds could degrade groundwater quality. However, groundwater in salt marshes is naturally brackish and thus not potable. Consequently, the human use of such groundwater is limited to industrial purposes.

The Category 2 surface water and groundwater impact issues identified in the GEIS are listed in the following table.

Table B-1. Summary of Findings on NEPA Issues for License Renewal of Nuclear Power Plants		
Issue	**Category**	**Finding**
Surface Water Resources		
Surface water use conflicts (plants with cooling ponds or cooling towers using makeup water from a river)	2	SMALL or MODERATE. Impacts could be of small or moderate significance, depending on makeup water requirements, water availability, and competing water demands.
Groundwater Resources		
Groundwater use conflicts (plants that withdraw more than 100 gallons per minute [gpm])	2	SMALL, MODERATE, or LARGE. Plants that withdraw more than 100 gpm could cause groundwater use conflicts with nearby groundwater users.
Groundwater use conflicts (plants with closed-cycle cooling systems that withdraw makeup water from a river)	2	SMALL, MODERATE, or LARGE. Water use conflicts could result from water withdrawals from rivers during low-flow conditions, which may affect aquifer recharge. The significance of impacts would depend on makeup water requirements, water availability, and competing water demands.
Groundwater quality degradation (plants with cooling ponds at inland sites)	2	SMALL, MODERATE, or LARGE. Inland sites with closed-cycle cooling ponds could degrade groundwater quality. The significance of the impact would depend on cooling pond water quality, site hydrogeologic conditions (including the interaction of surface water and groundwater), and the location, depth, and pump rate of water wells.
Radionuclides released to groundwater	2	SMALL or MODERATE. Leaks of radioactive liquids from plant components and pipes have occurred at numerous plants. Groundwater protection programs have been established at all operating nuclear power plants to minimize the potential impact from any inadvertent releases. The magnitude of impacts would depend on site-specific characteristics.

2. Determine if there is new information on these issues that should be evaluated. The following sources of information should be included in the search for new information:

- The applicant's ER. An applicant is required by 10 CFR 51.53(c)(3)(iv) to disclose new and significant information regarding the environmental impacts of license renewal of which it is aware. In reviewing the applicant's ER, consider the applicant's process for discovering new information and evaluating the significance of any new information discovered.

- Records of public scoping meetings and correspondence related to the application. Compare information presented by the public with information considered in the GEIS.

- Identify relative sources of information used for evaluating impacts, including:

    - Studies and monitoring programs: Briefly summarize any studies or monitoring programs that provide site-specific data and can assist with understanding the environmental impacts. Include the location, dates, objectives, methods, and results applicable to this license renewal application, and what data or data summaries might be available for NRC review.

If data are more than 5 years old, explain why the studies would or would not be relevant for assessing the effects of present and projected future plant operation over the term of license renewal. For example, show that both the potentially affected resources and the effect of the plant on them have remained and can be expected to remain unchanged over the term of license renewal.

- Communications with and views of regulatory agencies. Document any communications with regulatory agencies (e.g., EPA or other water quality permitting agencies) that are relevant to assessing impact and are not documented elsewhere in the environmental report. If relevant communications are documented elsewhere, refer the reader to the appropriate sections.

- Other sources. Give in-text citations to sources of data and information used to assess impact and provide a list of references at the end of the chapter.

3. Prepare a statement for the SEIS that:

- describes analysis of continued plant operations and refurbishment
- describes measures to mitigate adverse impacts
- provides the significance level of the environmental impacts
- describes new and significant information, if any

Additional specific guidance follows for each surface water and groundwater issue identified as plant-specific (Category 2) in the GEIS.

IV. EVALUATION FINDINGS

The depth and extent of the input to the SEIS would be determined by the analysis required to reach a conclusion related to the potential surface and groundwater impacts from continued plant operations during the license renewal term and refurbishment. The information that should be included in the SEIS is described in the review procedures.

V. IMPLEMENTATION

The method described in this ESRP would be used in evaluating conformance with the Commission's regulations, except in those cases in which the applicant for license renewal proposes an acceptable alternative for complying with specified portions of the regulations.

VI. BIBLIOGRAPHY

10 CFR Part 51. *Code of Federal Regulations*, Title 10, *Energy*, Part 51, "Environmental Protection Regulations for Domestic Licensing and Related Regulatory Functions."

40 CFR Part 6. *Code of Federal Regulations*, Title 40, *Protection of Environment*, Part 6, "Procedures for Implementing the National Environmental Policy Act and Assessing the Environmental Effects Abroad of EPA Actions."

40 CFR Part 122. *Code of Federal Regulations*, Title 40, *Protection of Environment*, Part 122, "EPA Administered Permit Programs: The National Pollutant Discharge Elimination System."

40 CFR Part 124. *Code of Federal Regulations*, Title 40, *Protection of Environment*, Part 124, "Procedures for Decisionmaking."

40 CFR Part 125. *Code of Federal Regulations*, Title 40, *Protection of Environment*, Part 125, "Criteria and Standards for the National Pollutant Discharge Elimination System."

40 CFR Part 133. *Code of Federal Regulations*, Title 40, *Protection of Environment*, Part 133, "Secondary Treatment Regulation."

40 CFR Part 149. *Code of Federal Regulations*, Title 40, *Protection of Environment*, Part 149, "Sole Source Aquifers."

40 CFR Part 165. *Code of Federal Regulations*, Title 40, *Protection of Environment*, Part 165, "Pesticide Management and Disposal."

40 CFR Part 403. *Code of Federal Regulations*, Title 40, *Protection of Environment*, Part 403, "General Pretreatment Regulations for Existing and New Sources of Pollution."

40 CFR Part 423. *Code of Federal Regulations*, Title 40, *Protection of Environment*, Part 423, "Steam Electric Power Generating Point Source Category."

40 CFR Part 700. *Code of Federal Regulations*, Title 40, *Protection of Environment*, Part 700, "General."

40 CFR Part 702. *Code of Federal Regulations*, Title 40, *Protection of Environment*, Part 702, "General Practices and Procedures."

40 CFR Part 704. *Code of Federal Regulations*, Title 40, *Protection of Environment*, Part 704, "Reporting and Recordkeeping Requirements."

40 CFR Part 707. *Code of Federal Regulations*, Title 40, *Protection of Environment*, Part 707, "Chemical Imports and Exports."

40 CFR Part 710. *Code of Federal Regulations*, Title 40, *Protection of Environment*, Part 710, "Compilation of the TSCA Chemical Substance Inventory."

40 CFR Part 711. *Code of Federal Regulations*, Title 40, *Protection of Environment*, Part 711, "TSCA Chemical Data Reporting Requirements."

40 CFR Part 712. *Code of Federal Regulations*, Title 40, *Protection of Environment*, Part 712, "Chemical Information Rules."

40 CFR Part 716. *Code of Federal Regulations*, Title 40, *Protection of Environment*, Part 716, "Health and Safety Data Reporting."

Federal Water Pollution Control Act (FWPCA), as amended. 33 USC 1251 et seq. (i.e., Clean Water Act).

PUD No. 1 of Jefferson County v. Washington Department of Ecology, 511 U.S. 700 (1994). National Environmental Policy Act of 1969 (NEPA). 42 USC 4321 et seq.

U.S. Nuclear Regulatory Commission (NRC). 2013. *Generic Environmental Impact Statement* (GEIS) *for License Renewal of Nuclear Plants*. NUREG-1437, Vols. 1, 2, and 3, Revision 1. Office of Nuclear Reactor Regulation, Washington, D.C.

U.S. NUCLEAR REGULATORY COMMISSION
ENVIRONMENTAL STANDARD REVIEW PLAN
OFFICE OF NUCLEAR REACTOR REGULATION

4.4.1 SURFACE WATER AND GROUNDWATER USE CONFLICTS

I. AREAS OF REVIEW

This environmental standard review plan (ESRP) provides guidance for the review of impacts on surface water and groundwater use from continued plant operations during the renewal term and refurbishment. Impacts are discussed in the *Generic Environmental Impact Statement for License Renewal of Nuclear Plants* (GEIS; NUREG-1437, Volumes 1, 2, and 3, Revision 1).

The scope includes (1) review of the discussion of surface water and groundwater use conflicts in the GEIS, (2) review of the applicant's environmental report (ER), (3) identifying and addressing any new and significant information, and (4) preparing input to the supplemental environmental impact statement (SEIS).

Data and Information Needs

The types of data and information needed would be affected by nuclear power plant site- and plant-specific factors. The following data or information may be needed:

- the applicant's ER

- the GEIS

- new information on surface water and groundwater use identified by the public and other information sources

II. ACCEPTANCE CRITERIA

Acceptance criteria for the evaluation of surface water and groundwater-use are addressed in ESRP Section 4.4, Water Resources.

Technical Rationale

The review conducted under this ESRP leads to the preparation of SEIS sections that incorporate the conclusions in the GEIS related to surface water and groundwater-use impacts of continued plant operations during the license renewal term, refurbishment, and alternatives to the proposed action. The review should also address any new and significant information.

III. REVIEW PROCEDURES

Suggested steps for the review process are as follows:

1. Review the discussion of surface water and groundwater-use issues in the GEIS to identify the information considered and the conclusions reached. This step establishes the basis for evaluating information identified by the applicant, the public, and the staff.

2. Use the Review Procedures in ESRP Section 4.4, Water Resources, with the following additions:

    -   a description of the applicant's process for identifying new and potentially significant information

    -   any new information included in the ER on the groundwater-use and quality issues known to the applicant and the public

    -   any currently employed or proposed practices and measures to control or limit operational water-use impact

    -   summary of statutory and other legal restrictions relating to water use or specific water-body restrictions on water use imposed by State or Federal regulations

    -   Federal, State, regional, local, and affected American Indian Tribal standards and regulations applicable to water quality and water use; determine if there are any State policies regarding hydraulic continuity

    -   proposed means to ensure operational compliance with water-quality and water-use standards and regulations

3. Prepare a statement for the SEIS that:

    -   describes analysis of continued plant operations and refurbishment

    -   describes measures to mitigate adverse impacts

    -   provides the significance level of the environmental impacts

- describes new and significant information, if any

IV. EVALUATION FINDINGS

The depth and extent of the input to the SEIS would be determined by the analysis required to reach a conclusion related to the potential surface water and groundwater-use conflicts from continued plant operations during the license renewal term and refurbishment. The information that should be included in the SEIS is described in the review procedures.

V. IMPLEMENTATION

The method described in this ESRP would be used in evaluating conformance with the Commission's regulations, except in those cases in which the applicant for license renewal proposes an acceptable alternative for complying with specified portions of the regulations.

VI. BIBLIOGRAPHY

The bibliography for this section is provided in ESRP Section 4.4, Water Resources.

ENVIRONMENTAL STANDARD REVIEW PLAN

OFFICE OF NUCLEAR REACTOR REGULATION

4.4.2 GROUNDWATER USE CONFLICTS (PLANTS THAT WITHDRAW MORE THAN 100 GPM)

I. AREAS OF REVIEW

This environmental standard review plan (ESRP) provides guidance for the review of the potential groundwater-use conflicts at plants pumping more than 100 gpm for potable and service water and operational dewatering, as well as those using Ranney wells. Impacts are discussed in the *Generic Environmental Impact Statement for License Renewal of Nuclear Plants* (GEIS; NUREG-1437, Volumes 1, 2, and 3, Revision 1).

The scope includes (1) review of the discussion of groundwater use conflicts in the GEIS, (2) review the applicant's environmental report (ER), (3) identifying and addressing any new and significant information, and (4) preparing input to the supplemental environmental impact statement (SEIS).

Data and Information Needs

The types of data and information needed would be affected by nuclear power plant site- and plant-specific factors. The following data or information may be needed:

- the applicant's ER

- the GEIS

- new information on groundwater-use conflicts identified by the public and other information

- sources

4.4.2-1

II. ACCEPTANCE CRITERIA

Acceptance criteria for the evaluation of groundwater-use conflicts are addressed in ESRP Section 4.4, Water Resources.

<u>Technical Rationale</u>

The review conducted under this ESRP leads to the preparation of SEIS sections that incorporate the conclusions in the GEIS related to groundwater-use conflicts from continued plant operations during the license renewal term, refurbishment, and alternatives to the proposed action. The review should also address any new and significant information.

III. REVIEW PROCEDURES

Suggested steps for the review process are as follows:

1. Review the discussion of the potential for groundwater water-use conflicts with nearby groundwater users at plants pumping more than 100 gpm for potable and service water and operational dewatering in the GEIS. This step establishes the basis for evaluating information identified by the applicant, the public, and the staff.

2. Determine the seasonal groundwater pumpage needs for the plant. If any season has an average groundwater pumpage of greater than 100 gpm, then continue the analysis at Step 3. Otherwise, prepare a statement for the SEIS that describes the plant's groundwater use and concludes that there are no impacts resulting from groundwater pumpage for potable and service water and operational dewatering.

3. Determine the extent of the influence of the plant's well(s) predicted by either standard analytic approaches or numerical models. Steady-state analytic approaches can be used with the maximum seasonal pumping rates. Numerical models can be used either with the maximum pumping rate to estimate steady-state drawdown or with the average seasonal pumping rates for a transient simulation of the drawdown. Model results should be validated with any piezometer observations. Possible impacts on predictions from heterogeneous aquifer parameters, particularly stratigraphy, should be considered. If the extent of the cone of depression caused by the plant's well(s) extends beyond the site's boundary, then continue the analysis.

4. Determine the magnitude of the reduction in yield resulting from the plant's pumpage predicted by numerical procedures. If the drawdown extends beyond the site boundary and into a zone influenced by other wells, then continue the analysis.

5. Use the Review Procedures in ESRP Section 4.4, Water Resources, with the following additions:

    *   descriptions of the site and local groundwater aquifers including geohydrologic characterization data

- descriptions of the spatial and seasonal changes in water table elevation and pumpage rates for wells both inside and outside the site boundary

- descriptions of any currently employed or proposed practices and measures to control or limit operational water-use impacts

- descriptions of Federal, State, regional, local, and affected American Indian Tribal agencies' standards and regulations applicable to groundwater use

- descriptions of proposed means to ensure operational compliance with water-quality and water-use standards and regulations.

6. Review the applicant's ER, including:

- the applicant's process for identifying new and potentially significant information

- any new information included in the ER on the groundwater-use and quality issues known to the applicant or the public

- any currently employed or proposed practices and measures to control or limit operational water-use impact

- summary of statutory and other legal restrictions relating to water use or specific water-body restrictions on water use imposed by State or Federal regulations

- proposed means to ensure operational compliance with water-quality and water-use standards and regulations

7. Prepare a statement for the SEIS that:

- describes analysis of continued plant operations and refurbishment

- describes measures to mitigate adverse impacts

- provides the significance level of the environmental impacts

- describes new and significant information, if any

IV. EVALUATION FINDINGS

The depth and extent of the input to the SEIS would be determined by the analysis required to reach a conclusion related to the potential groundwater-use conflicts from continued plant operations during the license renewal term and refurbishment. The information that should be included in the SEIS is described in the review procedures.

V. IMPLEMENTATION

The method described in this ESRP would be used in evaluating conformance with the Commission's regulations, except in those cases in which the applicant for license renewal proposes an acceptable alternative for complying with specified portions of the regulations.

VI. BIBLIOGRAPHY

The bibliography for this section is provided in ESRP Section 4.4, Water Resources.

ENVIRONMENTAL STANDARD
REVIEW PLAN

4.4.3 GROUNDWATER USE CONFLICTS (PLANTS WITH CLOSED-CYCLE COOLING SYSTEMS THAT WITHDRAW MAKEUP WATER FROM A RIVER)

I. AREAS OF REVIEW

This environmental standard review plan (ESRP) provides guidance for the review of groundwater-use conflicts resulting from surface-water withdrawals from a river during low-flow conditions. Impacts are discussed in the *Generic Environmental Impact Statement* (GEIS) *for License Renewal of Nuclear Plants* (NUREG-1437, Volumes 1, 2, and 3, Revision 1).

The scope includes (1) review of the discussion of groundwater use conflicts in the GEIS, (2) review the applicant's environmental report (ER), (3) identifying and addressing any new and significant information, and (4) preparing input to the supplemental environmental impact statement (SEIS).

Data and Information Needs

The types of data and information needed would be affected by nuclear power plant site- and plant-specific factors. The following data or information may be needed:

- the applicant's ER

- the GEIS

- new information on groundwater-use conflicts identified by the public and other information sources

II. ACCEPTANCE CRITERIA

Acceptance criteria for the evaluation of groundwater-use conflicts are addressed in ESRP Section 4.4, Water Resources.

Technical Rationale

The review conducted under this ESRP leads to the preparation of SEIS sections that incorporate the conclusions in the GEIS related to groundwater-use conflicts from continued plant operations during the license renewal term, refurbishment, and alternatives to the proposed action. The review should also address any new and significant information.

III. REVIEW PROCEDURES

Suggested steps for the review process are as follows:

1. Review the discussion in the GEIS of potential groundwater water-use conflicts resulting from surface-water withdrawals from small water bodies during low-flow conditions that may affect aquifer recharge. This step establishes the basis for evaluating information identified by the applicant, the public, and the staff.

2. Determine whether the river used for makeup water supply is oversubscribed (i.e., the demand for water exceeds supply) during any season. Water-use permits often include specific restrictions on withdrawals during certain low-flow conditions. If the basin is oversubscribed, continue the analysis.

3. Determine whether the river recharges the aquifer or the aquifer discharges into the river. If the aquifer consistently discharges to the river, then groundwater withdrawals would not be impacted by changes in river flow, whereas the river flows would be impacted by the groundwater withdrawals, although often not significantly. If the aquifer is consistently recharged by the river, then ground-water withdrawals would be impacted by changes in river flow, whereas the river flow would not be significantly impacted by the groundwater withdrawals. Often the direction of water transfer between rivers and their associated aquifers alternates back and forth as one moves downstream. By comparing the piezometer data from the affected aquifer with the river stage height data, the direction of flow can be determined. If the aquifer does not consistently discharge into the river downstream from the makeup water withdrawal location, continue the analysis.

4. Determine the magnitude of the reduction in groundwater yield resulting from the plant's cooling tower makeup water withdrawal. Estimating the magnitude of the reduction of groundwater yield generally requires application of analytic or numerical models. Only those wells located in areas downstream from the makeup water diversion whose contributing area includes recharge from the river need be considered. Sensitivity analyses should be included on the parameters governing the

exchange of water between the river and the aquifer. Based on the magnitude of the reduction in yield, the impact would be SMALL, MODERATE, or LARGE.

5. Use the Review Procedures in ESRP Section 4.4, Water Resources, with the following additions:

    - descriptions of the site, the affected river, and the local groundwater aquifers, including geohydrologic characterization data

    - the spatial and seasonal changes in water table elevation, surface withdrawals, groundwater withdrawals, stream stage height for the river, and the aquifer in hydraulic connection to the river

    - any currently employed or proposed practices and measures to control or limit operational water-use impacts

    - Federal, State, regional, local, and affected American Indian Tribal standards and regulations applicable to groundwater and surface-water use

    - proposed means to ensure operational compliance with water-use permits, standards, and regulations

6. Review the applicant's ER, including:

    - the applicant's process for identifying new and potentially significant information

    - any new information included in the ER on the groundwater-use and quality issues known to the applicant or the public

    - any currently employed or proposed practices and measures to control or limit operational water-use impact

    - summary of statutory and other legal restrictions relating to water use or specific water-body restrictions on water use imposed by State or Federal regulations

    - proposed means to ensure operational compliance with water-quality and water-use standards and regulations

7. Prepare a statement for the SEIS that:

    - describes analysis of continued plant operations and refurbishment

    - describes measures to mitigate adverse impacts

    - provides the significance level of the environmental impacts

    - describes new and significant information, if any

IV. EVALUATION FINDINGS

The depth and extent of the input to the SEIS would be determined by the analysis required to reach a conclusion related to the potential groundwater-use conflicts from continued plant operations during the license renewal term and refurbishment. The information that should be included in the SEIS is described in the review procedures.

V. IMPLEMENTATION

The method described in this ESRP would be used in evaluating conformance with the Commission's regulations, except in those cases in which the applicant for license renewal proposes an acceptable alternative for complying with specified portions of the regulations.

VI. BIBLIOGRAPHY

The bibliography for this section is provided in ESRP Section 4.4, Water Resources.

U.S. NUCLEAR REGULATORY COMMISSION

ENVIRONMENTAL STANDARD
REVIEW PLAN

OFFICE OF NUCLEAR REACTOR REGULATION

4.4.4 GROUNDWATER QUALITY DEGRADATION (PLANTS WITH COOLING PONDS AT INLAND SITES)

I. AREAS OF REVIEW

This environmental standard review plan (ESRP) provides guidance for the review of the potential impact of groundwater quality degradation resulting from closed cycle cooling ponds at inland sites. Impacts are discussed in the *Generic Environmental Impact Statement for License Renewal of Nuclear Plants* (GEIS; NUREG-1437, Volumes 1, 2, and 3, Revision 1).

The scope includes (1) review of the discussion of groundwater quality degradation in the GEIS, (2) review the applicant's environmental report (ER), (3) identifying and addressing any new and significant information, and (4) preparing input to the supplemental environmental impact statement (SEIS).

Data and Information Needs

The types of data and information needed would be affected by nuclear power plant site- and plant-specific factors. The following data or information may be needed:

- the applicant's ER

- the GEIS

- new information on groundwater quality degradation identified by the public and other information sources

---

USNRC ENVIRONMENTAL STANDARD REVIEW PLAN

Environmental standard review plans are prepared for the guidance of the Office of Nuclear Reactor Regulation staff responsible for environmental reviews for nuclear power plants. These documents are made available to the public as part of the Commission's policy to inform the nuclear industry and the general public of regulatory procedures and policies. Environmental standard review plans are not substitutes for regulatory guides or the Commission's regulations and compliance with them is not required. These supplemental environmental standard review plans are keyed to Regulatory Guide 4.2, Supplement 1, "Preparation of Environmental Reports for Nuclear Power Plant License Renewal Applications."

Published environmental standard review plans will be revised periodically, as appropriate, to accommodate comments and to reflect new information and experience.

Comments and suggestions for improvement will be considered and should be sent to the U.S. Nuclear Regulatory Commission, Office of Nuclear Reactor Regulation, Washington, DC 20555-0001.

II. ACCEPTANCE CRITERIA

Acceptance criteria for the evaluation of groundwater quality degradation are addressed in ESRP Section 4.4, Water Resources.

Technical Rationale

The review conducted under this ESRP leads to the preparation of SEIS sections that incorporate the conclusions in the GEIS related to groundwater quality degradation impacts of continued plant operations during the license renewal term, refurbishment, and alternatives to the proposed action. The review should also address any new and significant information.

III. REVIEW PROCEDURES

Suggested steps for the review process are as follows:

1. Review the discussion of groundwater quality degradation resulting from closed-cycle cooling-pond sites leaking into aquifers beneath inland sites in the GEIS. This step establishes the basis for evaluating information identified by the applicant, the public, and the staff.

2. Determine the evolving chemical composition of the cooling pond water. Closed-cycle cooling ponds have concentrations of total dissolved solids, heavy metals, and chlorinated organic compounds as a result of evaporation, contact with plant equipment, and water-treatment systems, respectively. These concentrations can evolve over time. The current chemical composition of the cooling water should be described, as well as the estimated chemical composition throughout the renewal term.

3. Review monitoring data on the chemical composition of groundwater in the vadose zone and aquifer that would likely receive water infiltrating from the cooling pond, as well as groundwater unaffected by the cooling pond. If the ambient groundwater quality in the aquifer is better than the estimated quality of the cooling pond water during the license renewal term, then continue with the analysis.

4. Review monitoring data on the infiltration from the cooling ponds to the water table. If the cooling ponds have no liners or the liners are not expected to remain impermeable throughout the license renewal term, then continue with the analysis.

5. Describe the estimated infiltration rate from the ponds throughout the license renewal term. These estimates should be used as the boundary conditions for a groundwater flow and transport model. Vadose zone transport can be neglected if the water infiltrating beneath the cooling pond is assumed to immediately enter the aquifer. If the predicted groundwater plume associated with a conservative nonsorbing tracer is likely to enter the zone of influence of a well, then continue the analysis.

6. Describe the changes in water quality for each of the impacted wells. Both the timing and magnitude of water quality changes should be described. Because this analysis would require the

application of groundwater flow and transport simulation models, describe the model calibration activities and any peer-review activities. Compare the predicted changes in groundwater quality to the uses for which the groundwater is needed to assess the magnitude of the impact.

7. Use the Review Procedures in ESRP Section 4.4, Water Resources, with the following additions:

   - cooling pond characteristics (e.g., use of liners, use of impermeable materials, impermeable soils) that would retard/prevent infiltration into local aquifers

   - types and concentrations of impurities in the cooling pond water and chemistry of soils along pathways to local aquifers to determine whether cooling pond water can contaminate the groundwater

   - quality of water of local aquifers that could be affected by infiltration of cooling-pond water

   - Federal, State, regional, local, and affected American Indian Tribal agencies' groundwater quality requirements with emphasis on any changes to these requirements that have occurred during the plant's initial license term and any anticipated changes to those requirements during the license renewal term

   - offsite groundwater users who could be affected by the degradation of aquifers; characterization should include locations and elevations of offsite wells, their pumping rates, and the water needs of groundwater users

   - the predicted cumulative effects of using closed cycle cooling ponds on groundwater quality. This description should include maps of the contamination plume. Information should be provided on groundwater contamination existing at the time of license renewal application and projected contamination during the license renewal term

   - the mitigation measures proposed to prevent or minimize groundwater quality degradation and the estimated impact of implementing these measures. Explain the reasons for not implementing any measures that were considered but rejected

8. Review the applicant's ER, including:

   - the applicant's process for identifying new and potentially significant information

   - any new information included in the ER on the groundwater quality degradation issues known to the applicant and the public

   - any currently employed or proposed practices and measures to control or limit operational water-use impact

   - summary of statutory and other legal restrictions relating to water use or specific water-body restrictions on water use imposed by State or Federal regulations

   - proposed means to ensure operational compliance with water-quality and water-use standards and regulations

9. Prepare a statement for the SEIS that:

- describes analysis of continued plant operations and refurbishment
- describes measures to mitigate adverse impacts
- provides the significance level of the environmental impacts
- describes new and significant information, if any

IV. EVALUATION FINDINGS

The depth and extent of the input to the SEIS would be determined by the analysis required to reach a conclusion related to the potential groundwater quality degradation from continued plant operations during the license renewal term and refurbishment. The information that should be included in the SEIS is described in the review procedures.

V. IMPLEMENTATION

The method described in this ESRP would be used in evaluating conformance with the Commission's regulations, except in those cases in which the applicant for license renewal proposes an acceptable alternative for complying with specified portions of the regulations.

VI. BIBLIOGRAPHY

The bibliography for this section is provided in ESRP Section 4.4, Water Resources.

4.4.5 RADIONUCLIDES RELEASED TO GROUNDWATER

I. AREAS OF REVIEW

This environmental standard review plan (ESRP) provides guidance for the review of the potential for radionuclides released to groundwater as a result of continued plant operations during the renewal term and refurbishment. Impacts are discussed in the *Generic Environmental Impact Statement for License Renewal of Nuclear Plants* (GEIS; NUREG-1437, Volumes 1, 2, and 3, Revision 1).

The scope includes (1) review of the discussion of radionuclides released to groundwater in the GEIS, (2) review the applicant's environmental report (ER), (3) identifying and addressing any new and significant information, and (4) preparing input to the supplemental environmental impact statement (SEIS).

Data and Information Needs

The types of data and information needed would be affected by nuclear power plant site- and plant-specific factors. The following data or information may be needed:

- the applicant's ER

- the GEIS

- new information on radionuclides released to groundwater identified by the public and other information sources

II. ACCEPTANCE CRITERIA

Acceptance criteria for the evaluation of radionuclides released to groundwater are addressed in ESRP Section 4.4, Water Resources.

Technical Rationale

The review conducted under this ESRP leads to the preparation of SEIS sections that incorporate the conclusions in the GEIS related to radionuclides released to groundwater impacts of continued plant operations during the license renewal term, refurbishment, and alternatives to the proposed action. The review should also address any new and significant information.

III. REVIEW PROCEDURES

Suggested steps for the review process are as follows:

1. Review the discussion of the potential for radionuclides released to groundwater in the GEIS. This step establishes the basis for evaluating information identified by the applicant, the public, and the staff.

2. Use the Review Procedures in ESRP Section 4.4, Water Resources, with the following additions: For plants that have groundwater monitoring systems with wells, review and describe:

   - the locations of monitoring wells and water supply wells, including construction information such as depth, diameter, screened interval, and construction material

   - depths of wells and water table elevations

   - description of groundwater flow for each aquifer or hydrostratigraphic unit beneath the site

   - description of radionuclide concentrations across the site (e.g, tritium concentrations in pCi/L)

   - for plants that rely on a system other than a groundwater monitoring system composed of wells, describe the program used for preventing, detecting, and responding to inadvertent releases of radioactive materials into the groundwater

   - describe the plant's groundwater protection program

3. Review the applicant's ER, including:

   - the applicant's process for identifying new and potentially significant information

   - any new information included in the ER on the radionuclides released to groundwater issues known to the applicant or the public

   - any currently employed or proposed practices and measures to control or limit operational water-use impact (best management practices)

- summary of statutory and other legal restrictions relating to water use or specific water-body restrictions on water use imposed by State or Federal regulations

- proposed means to ensure operational compliance with water-quality and water-use standards and regulations

4. Prepare a statement for the SEIS that:

- describes analysis of continued plant operations and refurbishment

- describes measures to mitigate adverse impacts, if any

- provides the significance level of the environmental impacts, if any

- describes new and significant information, if any

IV. EVALUATION FINDINGS

The depth and extent of the input to the SEIS would be determined by the analysis required to reach a conclusion related to the potential radionuclides released to groundwater from continued plant operations during the license renewal term and refurbishment. The information that should be included in the SEIS is described in the review procedures.

V. IMPLEMENTATION

The method described in this ESRP would be used in evaluating conformance with the Commission's regulations, except in those cases in which the applicant for license renewal proposes an acceptable alternative for complying with specified portions of the regulations.

VI. BIBLIOGRAPHY

The bibliography for this section is provided in ESRP Section 4.4, Water Resources.

ENVIRONMENTAL STANDARD
REVIEW PLAN

OFFICE OF NUCLEAR REACTOR REGULATION

4.5 ECOLOGICAL RESOURCES

I. AREAS OF REVIEW

This environmental standard review plan (ESRP) provides guidance for the review of ecological resource impacts from continued plant operations during the license renewal term and refurbishment. Impacts are discussed in the *Generic Environmental Impact Statement for License Renewal of Nuclear Plants* (GEIS; NUREG-1437, Volumes 1, 2, and 3, Revision 1).

The scope includes (1) review of the discussion of ecological resource impacts in the GEIS, (2) review of the applicant's environmental report (ER), (3) identifying and addressing any new and significant information, and (4) preparing input to the supplemental environmental impact statement (SEIS).

Ecological impact assessment for license renewal is different from that for original licensing, which is predictive or prospective in that it assumes a proposed stressor and proceeds to estimate impacts. License renewal, which occurs after the plant has a history of operation, can combine both prospective assessment to predict impacts during the license renewal term and retrospective assessment to assess the observed effects of past operation. For example, ecological modeling could be used to predict future impacts (in either original licensing or license renewal), while empirical statistical analysis could be used to assess past impacts based on actual observations (in license renewal only). Suter (1993, Chapter 10) discusses the differences between prospective and retrospective assessment and appropriate techniques for their analysis.

USNRC ENVIRONMENTAL STANDARD REVIEW PLAN

Environmental standard review plans are prepared for the guidance of the Office of Nuclear Reactor Regulation staff responsible for environmental reviews for nuclear power plants. These documents are made available to the public as part of the Commission's policy to inform the nuclear industry and the general public of regulatory procedures and policies. Environmental standard review plans are not substitutes for regulatory guides or the Commission's regulations and compliance with them is not required. These supplemental environmental standard review plans are keyed to Regulatory Guide 4.2, Supplement 1, "Preparation of Environmental Reports for Nuclear Power Plant License Renewal Applications."

Published environmental standard review plans will be revised periodically, as appropriate, to accommodate comments and to reflect new information and experience.

Comments and suggestions for improvement will be considered and should be sent to the U.S. Nuclear Regulatory Commission, Office of Nuclear Reactor Regulation, Washington, DC 20555-0001.

Data and Information Needs

The types of data and information needed would be affected by nuclear power plant site- and plant-specific factors. The following data or information may be needed:

- the applicant's ER

- the GEIS

- new information on ecological resources identified by the public and other information sources

II. ACCEPTANCE CRITERIA (General for all ecological resource issues)

Acceptance criteria for the evaluation of ecological resource impacts are based on, but not limited to, the following regulations:

- 10 CFR 51.45(c), "Analysis." The environmental report must include an analysis that considers and balances the environmental effects of the proposed action, the environmental impacts of alternatives to the proposed action, and alternatives available for reducing or avoiding adverse environmental effects.

- 10 CFR 51.53(c)(2). The report must contain a description of the proposed action, including the applicant's plans to modify the facility or its administrative control procedures as described in accordance with 10 CFR 54.21 of this chapter. This report must describe in detail the affected environment around the plant, the modifications directly affecting the environment or any plant effluents, and any planned refurbishment activities. In addition, the applicant shall discuss in this report the environmental impacts of alternatives and any other matters discussed in 10 CFR 51.45.

- 10 CFR 51.53(c)(3)(ii)(A). If the applicant's plant utilizes cooling towers or cooling ponds and withdraws makeup water from a river, an assessment of the impact of the proposed action on water availability and competing water demands, the flow of the river, and related impacts on stream (aquatic) and riparian (terrestrial) ecological communities must be provided.

- 10 CFR 51.53(c)(3)(ii)(B). If the applicant's plant utilizes once-through cooling or cooling pond heat dissipation systems, the applicant shall provide a copy of current Clean Water Act (CWA) 316(b) determinations and, if necessary, a 316(a) variance in accordance with 40 CFR part 125, or equivalent State permits and supporting documentation. If the applicant cannot provide these documents, it shall assess the impact of the proposed action on fish and shellfish resources resulting from thermal changes and impingement and entrainment.

- 10 CFR 51.53(c)(3)(ii)((E). All license renewal applicants shall assess the impact of refurbishment, continued operations, and other license-renewal-related construction activities on important plant and animal habitats. Additionally, the applicant shall assess the impact of the proposed action on threatened or endangered species in accordance with Federal laws protecting wildlife, including but not limited to, the Endangered Species Act, and essential fish habitat in accordance with the Magnuson-Stevens Fishery Conservation and Management Act.

- 10 CFR 51.70(b). The draft environmental impact statement will be concise, clear, and analytic, and written in plain language with appropriate graphics.… The format provided in Section 1(a) of Appendix A of this subpart should be used. The NRC staff will independently evaluate and be responsible for the reliability of all information used in the draft environmental impact statement.

- 10 CFR 51.71(d), concerning the draft environmental impact statement, will include a preliminary analysis that considers and weighs the environmental effects of the proposed action, the environmental impacts of alternatives to the proposed action, and alternatives available for reducing or avoiding adverse environmental effects.

- 10 CFR 51.95(c), concerning renewal of an operating license or combined license for a nuclear power plant. Under Parts 52 or 54 of this chapter, the Commission shall prepare an environmental impact statement, which is a supplement to the Commission's NUREG-1437, "Generic Environmental Impact Statement for License Renewal of Nuclear Plants."

- 10 CFR Part 51, Appendix A to Subpart A, para. 7, concerning the environmental consequences of alternatives, including the proposed actions and any mitigating actions which may be taken. Alternatives eliminated from detailed study will be identified and a discussion of those alternatives will be confined to a brief statement of the reasons why the alternatives were eliminated. The level of information for each alternative considered in detail will reflect the depth of analysis required for sound decisionmaking.

- 10 CFR Part 51, Appendix B to Subpart A, "Environmental Effect of Renewing the Operating License of a Nuclear Power Plant," Table B-1, "Summary of Findings on NEPA Issues for License Renewal of Nuclear Power Plants"

- 40 CFR Part 122, concerning NPDES permit conditions specified in the CWA

- 40 CFR Part 423, concerning effluent guidelines and thermal standards

- Coastal Zone Management Act of 1972 (16 USC 1451 et seq.), requires that Federal actions, including licensing or permitting actions, be consistent with U.S. Department of Commerce (NOAA) approved State Coastal Management Plans, which protect resources in a state's coastal zone.

- Endangered Species Act of 1973 (16 USC 1531 et seq.), requires the identification of threatened and endangered species and critical habitats, and formal or informal consultation with the U.S. Fish and Wildlife Service and/or National Marine Fisheries Service.

- Bald and Golden Eagle Protection Act (16 USC 668 et seq.) prohibits the taking, possessing, selling, purchasing, bartering, transporting, importing, or exporting of, the bald or golden eagle, dead or alive, or any part, nest, or egg without a permit.

- Migratory Bird Treaty Act (16 USC 703, et seq.), concerning declaring that it is unlawful to take, import, export, possess, buy, sell, purchase, or barter any migratory bird. Feathers, or other parts of nests and eggs, and products made from migratory birds are also covered by the Act. "Take" is defined as pursuing, hunting, shooting, poisoning, wounding, killing, capturing, trapping, or collecting.

- Marine Mammal Protection Act of 1972 (16 USC 1361, et seq.), concerning the protection of marine mammals

- Marine Protection, Research, and Sanctuaries Act of 1972 (33 USC 1401, et seq.), concerning dumping of dredged material into the ocean

- Rivers and Harbors Appropriations Act of 1899 (33 USC 403), concerning the deposition of debris in navigable waters, or tributaries to such waters

- The Magnuson-Stevens Fishery Conservation and Management Act, as amended (16 USC 1801 et seq.), established the Essential Fish Habitat provisions to identify and protect important habitats of federally managed marine and anadromous fish species and shellfish species

- Clean Water Act (33 USC 1251 et seq., as amended), concerning restoration and maintenance of the chemical, physical, and biological integrity of water resources

Additional regulatory positions and specific criteria in support of the regulations identified above are as follows:

Compliance with environmental quality standards and requirements of the CWA is not a substitute for and does not negate the requirement for NRC to weigh the environmental impacts of the proposed action and to consider mitigation measures and alternatives to reduce adverse impacts. If the applicant provides a current NPDES permit, CWA 316(b) determination, or State permits and supporting documentation acceptable to the EPA at the time of license-renewal application, the NRC would consider them in its determination of the magnitude of the environmental impacts. When such assessments of impacts of impingement and entrainment of fish and shellfish in early life stages are not provided by the applicant, the NRC (possibly in conjunction with the permitting authority and other agencies having relevant expertise) would conduct its own assessment.

Technical Rationale

The review conducted under this ESRP leads to the preparation of SEIS sections that incorporate the conclusions in the GEIS related to ecological resource impacts of continued plant operations during the license renewal term, refurbishment, and alternatives to the proposed action. The review should also address any new and significant information.

III. REVIEW PROCEDURES (General for all ecological resource issues)

For all ecological issues, the same general approach can identify the environmental consequences of license renewal and its alternatives. This approach generally follows the framework for ecological risk assessment (EPA 1998).

1. Review the discussion of ecological resource impacts in the GEIS. This step establishes the basis for evaluating information identified by the applicant, the public, and the staff. Category 1 terrestrial and aquatic resource issues identified in the GEIS are listed in the following table.

Table B-1. Summary of Findings on NEPA Issues for License Renewal of Nuclear Power Plants		
Issue	Category	Finding
Terrestrial Resources		
Exposure of terrestrial organisms to radionuclides	1	SMALL. Doses to terrestrial organisms from continued operations and refurbishment associated with license renewal are expected to be well below exposure guidelines developed to protect these organisms.
Cooling system impacts on terrestrial resources (plants with once-through cooling systems or cooling ponds)	1	SMALL. No adverse effects to terrestrial plants or animals have been reported as a result of increased water temperatures, fogging, humidity, or reduced habitat quality. Due to the low concentrations of contaminants in cooling system effluents, uptake and accumulation of contaminants in the tissues of wildlife exposed to the contaminated water or aquatic food sources are not expected to be significant issues.
Cooling tower impacts on vegetation (plants with cooling towers)	1	SMALL. Impacts from salt drift, icing, fogging, or increased humidity associated with cooling tower operation have the potential to affect adjacent vegetation, but these impacts have been small at operating nuclear power plants and are not expected to change over the license renewal term.
Bird collisions with plant structures and transmission lines	1	SMALL. Bird collisions with cooling towers and other plant structures and transmission lines occur at rates that are unlikely to affect local or migratory populations and the rates are not expected to change.
Transmission line right-of-way (ROW) management impacts on terrestrial resources	1	SMALL. Continued ROW management during the license renewal term is expected to keep terrestrial communities in their current condition. Application of best management practices would reduce the potential for impacts.
Electromagnetic fields on flora and fauna (plants, agricultural crops, honeybees, wildlife, livestock)	1	SMALL. No significant impacts of electromagnetic fields on terrestrial flora and fauna have been identified. Such effects are not expected to be a problem during the license renewal term.
Aquatic Resources		
Impingement and entrainment of aquatic organisms (plants with cooling towers)	1	SMALL. Impingement and entrainment rates are lower at plants that use closed-cycle cooling with cooling towers because the rates and volumes of water withdrawal needed for makeup are minimized.

Table B-1. Summary of Findings on NEPA Issues for License Renewal of Nuclear Power Plants

Issue	Category	Finding
Entrainment of phytoplankton and zooplankton (all plants)	1	SMALL. Entrainment of phytoplankton and zooplankton has not been found to be a problem at operating nuclear power plants and is not expected to be a problem during the license renewal term.
Thermal impacts on aquatic organisms (plants with cooling towers)	1	SMALL. Thermal effects associated with plants that use cooling towers are expected to be small because of the reduced amount of heated discharge.
Infrequently reported thermal impacts (all plants)	1	SMALL. Continued operations during the license renewal term are expected to have small thermal impacts with respect to the following: - Cold shock has been satisfactorily mitigated at operating nuclear plants with once-through cooling systems, has not endangered fish populations or been found to be a problem at operating nuclear power plants with cooling towers or cooling ponds, and is not expected to be a problem. - Thermal plumes have not been found to be a problem at operating nuclear power plants and are not expected to be a problem. - Thermal discharge may have localized effects but is not expected to affect the larger geographical distribution of aquatic organisms. - Premature emergence has been found to be a localized effect at some operating nuclear power plants but has not been a problem and is not expected to be a problem. - Stimulation of nuisance organisms has been satisfactorily mitigated at the single nuclear power plant with a once-through cooling system where previously it was a problem. It has not been found to be a problem at operating nuclear power plants with cooling towers or cooling ponds and is not expected to be a problem.
Effects of cooling water discharge on dissolved oxygen, gas supersaturation, and eutrophication	1	SMALL. Gas supersaturation was a concern at a small number of operating nuclear power plants with once-through cooling systems but has been mitigated. Low dissolved oxygen was a concern at one nuclear power plant with a once-through cooling system but has been mitigated. Eutrophication (nutrient loading) and resulting effects on chemical and biological oxygen demands have not been found to be a problem at operating nuclear power plants.
Effects of nonradiological contaminants on aquatic organisms	1	SMALL. Best management practices and discharge limitations of NPDES permits are expected to minimize the potential for impacts to aquatic resources during continued operations and refurbishment associated with license renewal. Accumulation of metal contaminants has been a concern at a few nuclear power plants but has been satisfactorily mitigated by replacing copper alloy condenser tubes with those of another metal.
Exposure of aquatic organisms to radionuclides	1	SMALL. Doses to aquatic organisms are expected to be well below exposure guidelines developed to protect these aquatic organisms.

Table B-1. Summary of Findings on NEPA Issues for License Renewal of Nuclear Power Plants

Issue	Category	Finding
Effects of dredging on aquatic organisms	1	SMALL. Dredging at nuclear power plants is expected to occur infrequently, would be of relatively short duration, and would affect relatively small areas. Dredging is performed under permit from the U.S. Army Corps of Engineers, and possibly, from other State or local agencies.
Effects on aquatic resources (non-cooling-system impacts)	1	SMALL. Licensee application of appropriate mitigation measures is expected to result in no more than small changes to aquatic communities from their current condition.
Impacts of transmission line ROW management on aquatic resources	1	SMALL. Licensee application of best management practices to ROW maintenance is expected to result in no more than small impacts to aquatic resources.
Losses from predation, parasitism, and disease among organisms exposed to sublethal stresses	1	SMALL. These types of losses have not been found to be a problem at operating nuclear power plants and are not expected to be a problem during the license renewal term.

The Category 2 terrestrial and aquatic resources and special status species and habitats impact issues identified in the GEIS are listed in the following table.

Table B-1. Summary of Findings on NEPA Issues for License Renewal of Nuclear Power Plants

Issue	Category	Finding
Terrestrial Resources		
Effects on terrestrial resources (non-cooling system impacts)	2	SMALL, MODERATE, or LARGE. Impacts resulting from continued operations and refurbishment associated with license renewal may affect terrestrial communities. Application of best management practices would reduce the potential for impacts. The magnitude of impacts would depend on the nature of the activity, the status of the resources that could be affected, and the effectiveness of mitigation.
Water use conflicts with terrestrial resources (plants with cooling ponds or cooling towers using makeup water from a river)	2	SMALL or MODERATE. Impacts on terrestrial resources in riparian communities affected by water use conflicts could be of moderate significance.
Aquatic Resources		
Impingement and entrainment of aquatic organisms (plants with once-through cooling systems or cooling ponds)	2	SMALL, MODERATE, or LARGE. The impacts of impingement and entrainment are small at many plants, but may be moderate or even large at a few plants with once-through and cooling-pond cooling systems, depending on cooling system withdrawal rates and volumes and the aquatic resources at the site.
Thermal impacts on aquatic organisms (plants with once-through cooling systems or cooling ponds)	2	SMALL, MODERATE, or LARGE. Most of the effects associated with thermal discharges are localized and are not expected to affect overall stability of populations or resources. The magnitude of impacts, however, would depend on site-specific thermal plume characteristics and the nature of aquatic resources in the area.

Table B-1. Summary of Findings on NEPA Issues for License Renewal of Nuclear Power Plants		
Issue	**Category**	**Finding**
Water use conflicts with aquatic resources (plants with cooling ponds or cooling towers using makeup water from a river)	2	SMALL or MODERATE. Impacts on aquatic resources in stream communities affected by water use conflicts could be of moderate significance in some situations.
Special Status Species and Habitats		
Threatened, endangered, and protected species and essential fish habitat	2	The magnitude of impacts on threatened, endangered, and protected species, critical habitat, and essential fish habitat would depend on the occurrence of listed species and habitats and the effects of power plant systems on them. Consultation with appropriate agencies would be needed to determine whether special status species or habitats are present and whether they would be adversely affected by continued operations and refurbishment associated with license renewal.

2. Identify the relevant sources of information, which include:

- Studies and monitoring programs: Briefly summarize any studies that provide site-specific data and can help understand environmental impacts. Include the location, dates, objectives, the biological entities or attributes chosen for study, methods, and results applicable to this license renewal application, and what data or data summaries might be available for Nuclear Regulatory Commission (NRC) review.

  If data are more than 5 years old, explain why the studies would or would not be relevant for assessing the effects of present and projected future plant operation over the term of license renewal. For example, show that both the potentially affected resources and the effect of the plant on them have remained and can be expected to remain unchanged over the term of license renewal.

- Communications with and views of regulatory agencies: Document any communications with regulatory agencies (e.g., Environmental Protection Agency or other water quality permitting agencies) and resource agencies (e.g., National Marine Fisheries Service, U.S. Fish and Wildlife Service, State fish and wildlife agencies) that are relevant to assessing impacts and are not documented elsewhere in the environmental report. If relevant communications are documented elsewhere, refer the reader to the appropriate sections.

  Discuss the major points of view concerning environmental impacts of license renewal and an analysis of significant problems and objections raised by other Federal, State, and local agencies; affected Indian Tribes; and other interested stakeholders.

- Other sources: Give in-text citations to sources of data and information used to assess impacts and provide a literature cited section.

3. Identify specific biological resources and their attributes used for assessing impact. Because biological systems are complicated, only a subset of resources can be addressed.

- Identify potentially affected resource entities: Describe the potentially affected resources in terms of representative species, functional group of species (e.g., insectivores), communities, an ecosystem (e.g., oak-hickory forest), a specific valued habitat (e.g., wet meadows, submerged aquatic vegetation), unique places, or other entities of concern.

Additional guidance on identifying important species to be evaluated can be found in "U.S. Fish and Wildlife Service Mitigation Policy; Notice of Final Policy."[3] Contact Federal, State, and regional government agencies with jurisdiction over biological resources to assist with the identification of important species and habitats.

- Identify attributes of those resources potentially at risk: Identify the attributes of the resources of concern that are important to protect and potentially at risk (U.S. EPA 1998). If potentially adverse effects on a species, habitat, or other ecological resource are identified, the resource should be assessed concerning local, regional, and national social, economic, and ecological value.

  Biodiversity, which comprises the variation between and among biological entities and includes components of genetic, species, habitat, local ecosystem and regional ecosystem diversity, is an important biological attribute to consider (CEQ 1993).

4. Show the relationships (as in a causal link) between plant operation and the resource attributes. Lack of change in a resource attribute may not indicate lack of an adverse effect if plant operations have no link to resource levels. Relationships can be examined by identifying the pathways through which they act and expressing them as a risk hypotheses.

   If any adverse impacts are identified, describe mitigation measures that have been used to reduce the adverse impacts during the initial license term or that are expected to be used during the license renewal term and their expected effects. Briefly explain the rationale for not implementing any measures that were considered but rejected.

5. Assess and characterize impact. For each ecological issue, multiple natural populations or entities may be affected, and each population may have multiple measurable and susceptible attributes. The effects of power plant operation on any ecological attribute may be direct or indirect, and the assessment approach can be prospective or retrospective. With such complexity, examining a single line of evidence may not be sufficient to assess impact. In such cases, the reviewer should examine several lines of evidence involving several populations when data allow.

   Present in the SEIS a narrative explanation of the conceptual model for the analysis that includes the justification for selection of representative entities or their attributes, the hypothetical mechanism through which continued plant operation may act upon them, the approaches used to assess impact, and possible outcomes. If using multiple lines of evidence, explain in narrative the qualitative or quantitative method for combining them to make an overall assessment of impact. A typical

---

[3] The notice (46 FR 7644, January 23, 1981) establishes guidance for U.S. Fish and Wildlife Service personnel involved in making recommendations to protect or conserve fish and wildlife resources. Guidance is provided on the definition and identification of "evaluation species," evaluation of direct and indirect effects of a project on the evaluation species, on the levels of mitigation and on the various methods for accomplishing mitigation when adverse effects are identified. The types of species that should be considered are discussed in the notice at pages 7662 and 7663. In this environmental standard review plan, the terms "important species" and "evaluation species" are used interchangeably.

approach for accomplishing this in ecological risk assessment is to consider weight of evidence (e.g., Menzie et al. 1996, EPA 1998).

6. Review the applicant's ER, including:

   - the applicant's process for identifying new and potentially significant information

   - any new information included in the ER on ecological impact issues known to the applicant and the public

7. Prepare a statement for the SEIS that:

   - describes analysis of continued plant operations and refurbishment

   - describes measures to mitigate adverse impacts

   - provides the significance level of the environmental impacts

   - describes new and significant information, if any

IV. EVALUATION FINDINGS

The depth and extent of the input to the SEIS would be determined by the analysis required to reach a conclusion related to ecological resource impacts from continued plant operations during the license renewal term and refurbishment. The information that should be included in the SEIS is described in the review procedures.

V. IMPLEMENTATION

The method described in this ESRP would be used by the staff in evaluating conformance with the Commission's regulations, except in those cases in which the applicant for license renewal proposes an acceptable alternative for complying with specified portions of the regulations.

VI. BIBLIOGRAPHY

10 CFR Part 51. *Code of Federal Regulations*, Title 10, *Energy*, Part 51, "Environmental Protection Regulations for Domestic Licensing and Related Regulatory Functions."

40 CFR Part 122. *Code of Federal Regulations*, Title 40, *Protection of Environment*, Part 122, "EPA Administered Permit Programs: The National Pollutant Discharge Elimination System."

40 CFR Part 123. *Code of Federal Regulations*, Title 40, *Protection of Environment*, Part 123, "State Program Requirements."

46 FR 7644. U.S. Fish and Wildlife Service. Fish and Wildlife Service Mitigation Policy. 1981.

Bald and Golden Eagle Protection Act of 1940, as amended. 16 USC 668 et seq.

Council on Environmental Quality (CEQ). 1993. *Incorporating Biodiversity Considerations into Environmental Impact Analysis under the National Environmental Quality Act.* CEQ, Executive Office of the President, Washington, D.C.

Coastal Zone Management Act of 1972 (CZMA). 16 USC 1451 et seq.

Endangered Species Act. 16 USC 1531 et seq.

Federal Water Pollution Control Act of 1977 (Clean Water Act). 33 USC 1251, et seq.

Fish and Wildlife Coordination Act of 1958. 16 USC 661 et seq.

National Environmental Policy Act of 1969 (NEPA). 42 USC 4321 et seq.

Suter, G.W. 1993. "Chapter 10: Retrospective Risk Assessment." Pp. 311–364. In G.W. Suter, II (ed.). *Ecological Risk Assessment.* Lewis publishers, Boca Raton, Florida.

U.S. Environmental Protection Agency (EPA). 1998. *Guidelines for Ecological Risk Assessment.* EPA/630/R-95/002. Risk Assessment Forum, Washington, D.C.

U.S. Nuclear Regulatory Commission (NRC). 2013. *Generic Environmental Impact Statement for License Renewal of Nuclear Plants.* NUREG-1437, Vols. 1, 2, and 3, Revision 1. Office of Nuclear Reactor Regulation, Washington, D.C.

U.S. NUCLEAR REGULATORY COMMISSION
ENVIRONMENTAL STANDARD
REVIEW PLAN
OFFICE OF NUCLEAR REACTOR REGULATION

4.5.1 WATER USE CONFLICTS WITH ECOLOGICAL (TERRESTRIAL OR AQUATIC) RESOURCES (PLANTS WITH COOLING PONDS OR COOLING TOWERS USING MAKEUP WATER FROM A RIVER)

I. AREAS OF REVIEW

This environmental standard review plan (ESRP) provides guidance for the review of impacts of surface water use conflicts on ecological (terrestrial or aquatic) resources from continued plant operations during the license renewal term and refurbishment. Impacts are discussed in the *Generic Environmental Impact Statement for License Renewal of Nuclear Plants* (GEIS; NUREG-1437, Volumes 1, 2, and 3, Revision 1).

The scope includes (1) review of the discussion of ecological resource impacts in the GEIS, (2) review of the applicant's environmental report (ER), (3) identifying and addressing any new and significant information, and (4) preparing input to the supplemental environmental impact statement (SEIS). This ESRP section applies to plants with cooling ponds or cooling towers using makeup water from a river.

Data and Information Needs

The types of data and information needed would be affected by nuclear power plant site- and plant-specific factors. The following data or information may be needed:

- the applicant's ER

- the GEIS

USNRC ENVIRONMENTAL STANDARD REVIEW PLAN

- new information on surface water use conflicts on ecological resources identified by the public and other information sources

II. ACCEPTANCE CRITERIA

Acceptance criteria for the evaluation of impacts of surface water use conflicts on ecological resources are addressed in ESRP Section 4.5, Ecological Resources.

Technical Rationale

The review conducted under this ESRP leads to the preparation of SEIS sections that incorporate the conclusions in the GEIS related to impacts of surface water use conflicts on ecological resources from continued plant operations during the license renewal term, refurbishment, and alternatives to the proposed action. The review should also address any new and significant information.

III. REVIEW PROCEDURES

Suggested steps for the review process are as follows:

1. Review the discussion of surface water use conflicts on ecological resources in the GEIS to identify the information considered and the conclusions reached. This step establishes the basis for evaluating information identified by the applicant, the public, and the staff.

2. Follow the general approach for information and analysis content for all ecology issues, described in ESRP Section 4.5, Ecological Resources, with the following additions:

   - no additional surface water conflict information is needed for plants using once-through cooling systems or not specifically using cooling towers or cooling ponds

   - discussion of riparian impacts can be included in the discussion of terrestrial resources, and discussion of instream impacts can be included in the discussion of aquatic resources

3. Review the applicant's ER, including:

   - the applicant's process for identifying new and potentially significant information

   - any new information included in the ER on ecological impact issues known to the applicant and the public

4. Prepare a statement for the SEIS that:

   - describes analysis of continued plant operations and refurbishment

   - describes measures to mitigate adverse impacts

   - provides the significance level of the environmental impacts

   - describes new and significant information, if any

IV. EVALUATION FINDINGS

The depth and extent of the input to the SEIS would be determined by the analysis required to reach a conclusion related to ecological resource impacts from continued plant operations during the license renewal term and refurbishment. The information that should be included in the SEIS is described in the review procedures.

V. IMPLEMENTATION

The method described in this ESRP would be used by the staff in evaluating conformance with the Commission's regulations, except in those cases in which the applicant for license renewal proposes an acceptable alternative for complying with specified portions of the regulations.

VI. BIBLIOGRAPY

The bibliography for this section is provided in ESRP Section 4.5, Ecological Resources.

U.S. NUCLEAR REGULATORY COMMISSION

ENVIRONMENTAL STANDARD REVIEW PLAN

OFFICE OF NUCLEAR REACTOR REGULATION

4.5.2 EFFECTS ON TERRESTRIAL RESOURCES (NON-COOLING-SYSTEM IMPACTS)

I. AREAS OF REVIEW

This environmental standard review plan (ESRP) provides guidance for the review of effects on terrestrial resources from continued plant operations during the renewal term and refurbishment. Impacts are discussed in the *Generic Environmental Impact Statement for License Renewal of Nuclear Plants* (GEIS; NUREG-1437, Volumes 1, 2, and 3, Revision 1).

The scope includes (1) review of the discussion of impacts of continued plant operations and refurbishment on terrestrial ecosystems in the GEIS, (2) review of the applicant's environmental report (ER), (3) identifying and addressing any new and significant information, and (4) preparing input to the supplemental environmental impact statement (SEIS).

Data and Information Needs

The types of data and information needed would be affected by nuclear power plant site- and plant-specific factors. The following data or information may be needed:

- the applicant's ER

- the GEIS

- new information on terrestrial ecosystems identified by the public and other information sources

II. ACCEPTANCE CRITERIA

Acceptance criteria for the evaluation of impacts of continued plant operations and refurbishment on terrestrial ecosystems are addressed in ESRP Section 4.5, Ecological Resources.

Technical Rationale

The review conducted under this ESRP leads to the preparation of SEIS sections that incorporate the conclusions in the GEIS related to impacts of continued plant operations during the license renewal term and refurbishment on terrestrial ecosystems. This section should consider the impacts of alternatives to the proposed action. The review should also address any new and significant information.

III. REVIEW PROCEDURES

For all ecological issues, the same general approach can identify the environmental consequences of license renewal and its alternatives. This approach generally follows the framework for ecological risk assessment (EPA 1998).

1. Review the discussion of impacts of continued plant operations and refurbishment on terrestrial ecosystems in the GEIS. This step establishes the basis for evaluating information identified by the applicant, the public, and the staff.

2. Follow the general approach for information and analysis content for all ecology issues, described in ESRP Section 4.5, Ecological Resources, with the following additions:

   • Describe any activities associated with continued plant operations and refurbishment associated with license renewal that would involve disturbing any terrestrial habitat. If no area would be disturbed, note that fact and no further discussion of the issue is needed. Areas to be disturbed should be described in Chapter 3 of the SEIS concerning (1) the amount of land to be disturbed, (2) ecological characteristics of the habitat, (3) species of plants and animals found in the area, and (4) the extent to which the habitat is unusual. Note that the information and analysis for this issue overlaps the information and analysis for assessing impacts on threatened and endangered species.

   • In addition, if any license renewal activity would disturb any plant or wildlife habitat, describe any land that would be disturbed during transport and delivery of equipment, structures, or components; storage of materials; or construction associated with refurbishment. If any temporary or permanent structures would be built, provide a map of the site that includes the proposed location of these structures. If any road or bridge modifications would occur as a result of transport, describe potential effects to the terrestrial environment.

3. Review the applicant's ER, including:

- the applicant's process for identifying new and potentially significant information

- any new information included in the ER on ecological impact issues known to the applicant or the public

4. Prepare a statement for the SEIS that:

- describes analysis of continued plant operations and refurbishment

- describes measures to mitigate adverse impacts

- provides the significance level of the environmental impacts

- describes new and significant information, if any

IV. EVALUATION FINDINGS

The depth and extent of the input to the SEIS would be determined by the analysis required to reach a conclusion related to ecological resource impacts from continued plant operations during the license renewal term and refurbishment. The information that should be included in the SEIS is described in the review procedures.

V. IMPLEMENTATION

The method described in this ESRP would be used by the staff in evaluating conformance with the Commission's regulations, except in those cases in which the applicant for license renewal proposes an acceptable alternative for complying with specified portions of the regulations.

VI. BIBLIOGRAPHY

The bibliography for this section is provided in ESRP Section 4.5, Ecological Resources.

U.S. NUCLEAR REGULATORY COMMISSION

ENVIRONMENTAL STANDARD
REVIEW PLAN

OFFICE OF NUCLEAR REACTOR REGULATION

4.5.3 IMPINGEMENT AND ENTRAINMENT OF AQUATIC ORGANISMS (PLANTS WITH ONCE-THROUGH COOLING SYSTEMS OR COOLING PONDS)

I. AREAS OF REVIEW

This environmental standard review plan (ESRP) provides guidance for the review of impingement and entrainment impacts on aquatic organisms for plants with once-through cooling systems or cooling ponds from continued plant operations during the renewal term and refurbishment. Impacts are discussed in the *Generic Environmental Impact Statement for License Renewal of Nuclear Plants* (GEIS; NUREG-1437, Volumes 1, 2, and 3, Revision 1).

The scope includes (1) review of the discussion of impingement and entrainment impacts on aquatic organisms in the GEIS, (2) review of the applicant's environmental report (ER), (3) identifying and addressing any new and significant information, and (4) preparing input to the supplemental environmental impact statement (SEIS).

The impacts from cooling water intake are regulated through the National Pollutant Discharge Elimination System (NPDES) permit system. The Clean Water Act (CWA) requires that the location, design, construction, and capacity of the cooling water intake structure reflect the best technology available for minimizing environmental impacts. Responsibility for making this determination rests with the U.S. Environmental Protection Agency (EPA) or with its designees.

<u>Data and Information Needs</u>

The types of data and information needed would be affected by nuclear power plant site- and plant-specific factors. The following data or information may be needed:

- the applicant's ER

- the GEIS

- new information on impingement and entrainment of aquatic organisms identified by the public and other information sources

II. ACCEPTANCE CRITERIA

Acceptance criteria for the evaluation of impingement and entrainment of aquatic organisms are addressed in ESRP Section 4.5, Ecological Resources.

<u>Technical Rationale</u>

The review conducted under this ESRP leads to the preparation of SEIS sections that incorporate the conclusions in the GEIS related to impacts of impingement and entrainment of aquatic organisms from continued plant operations during the license renewal term, refurbishment, and alternatives to the proposed action. The review should also address any new and significant information.

III. REVIEW PROCEDURES

1. Review the discussion of impingement and entrainment of aquatic organisms in the GEIS. This step establishes the basis for evaluating information identified by the applicant, the public, and the staff.

2. Follow the general approach for information and analysis content for all ecology issues, described in ESRP Section 4.5, Ecological Resources, with the following additions:

   - The reviewer should obtain additional information about the plant's cooling water system design and operating procedures from the applicant's environmental report and other ESRP sections that describe components of the cooling system and the hydrodynamics and physical impacts of the intake. The following data or information may be needed to assess the impingement and entrainment impacts of aquatic organisms:

     - a copy of the plant's current NPDES permit and a 316(b) demonstration acceptable to the EPA or State permitting agency

     - a description of any proposed changes or refurbishment to the plant-cooling-system intake

- studies or monitoring programs of impingement and entrainment, including information on when and where conducted and by whom, the objective (why conducted), the biological entities or attributes chosen for study, methods and results (how conducted), and data summaries available for NRC review

- information on the potential for altered hydrodynamic characteristics that are induced by inlet system operation (e.g., altered circulation patterns) and may affect attraction, impingement, and entrainment of aquatic biota and determine the extent and seasonal variation of any such alterations

- information on the potential for the recirculation of heated effluent from the plant discharge system. If recirculation is predicted, the reviewer would need information to assess the impacts to impingement and entrainment.

3. Review the applicant's ER, including:

- the applicant's process for identifying new and potentially significant information

- any new information included in the ER on impingement and entrainment issues known to the applicant and the public

4. Prepare a statement for the SEIS that:

- describes analysis of continued plant operations and refurbishment

- summarizes the permitting documents

- states that a NPDES permit and 316(b) determination are current and available

- describes measures to mitigate adverse impacts

- provides the significance level of the environmental impacts

- describes new and significant information, if any

IV. EVALUATION FINDINGS

The depth and extent of the input to the SEIS would be determined by the analysis required to reach a conclusion related to impingement and entrainment of aquatic organisms from continued plant operations during the license renewal term and refurbishment. The information that should be included in the SEIS is described in the review procedures.

V. IMPLEMENTATION

The method described in this ESRP would be used by the staff in evaluating conformance with the Commission's regulations, except in those cases in which the applicant for license renewal proposes an acceptable alternative for complying with specified portions of the regulations.

VI. BIBLIOGRAPHY

The bibliography for this section is provided in ESRP Section 4.5, Ecological Resources.

4.5.4 THERMAL IMPACTS ON AQUATIC ORGANISMS (PLANTS WITH ONCE-THROUGH COOLING SYSTEMS OR COOLING PONDS)

I. AREAS OF REVIEW

This environmental standard review plan (ESRP) provides guidance for the review of thermal impacts on aquatic organisms for plants with once-through cooling systems or cooling ponds from continued plant operations during the renewal term and refurbishment. Impacts are discussed in the *Generic Environmental Impact Statement for License Renewal of Nuclear Plants* (GEIS; NUREG-1437, Volumes 1, 2, and 3, Revision 1).

The scope includes (1) review the discussion of thermal impacts on aquatic organisms in the GEIS, (2) review the applicant's environmental report (ER), (3) identify and address any new and significant information, and (4) prepare input to the supplemental environmental impact statement (SEIS).

The impacts from cooling water discharges are regulated through the National Pollutant Discharge Elimination System (NPDES) permit system. The Clean Water Act (CWA) requires that discharge system operation must ensure the protection and propagation of a balanced, indigenous population of shellfish, fish, and wildlife in and on the receiving water body. Responsibility for making this determination rests with the U.S. Environmental Protection Agency (EPA) or with its designees.

Discharge system impacts on aquatic biota may result from the effects of thermal alterations to the receiving water body. Major alterations are usually confined to a limited discharge area (the mixing zone), whereas lesser alterations may extend over a larger portion of the receiving water body. Adverse effects on biota that are transported through, migrate through, or are attracted to the mixing zone may be acute or chronic, and impacts may be reflected as changes in the populations of "important" species and in the structure and function of the ecosystem.

This thermal effects section represents a combination of issues in the GEIS. Thermal effects on natural species and communities have different manifestations that may overlap and grade into each other depending on thermal history of the receptor; the life stages, species, communities, or habitats exposed; environmental factors; refugia; temperatures, durations, and pattern of acclimation or acclimatization and of temperature alterations; inter alia. As a result, the source and exact form of thermal effects can be hard to differentiate in actual field studies. For example, if a field study indicates exclusion of an ecological resource in an area influenced by a thermal plume, the investigators would have difficulty determining if the mechanism is habitat change, migration of individuals away from the increased temperature (and to what fate), chronic effects including changes in timing of life history events, acute mortality, or any combination of these.

Data and Information Needs

The types of data and information needed would be affected by nuclear power plant site- and plant-specific factors. The following data or information may be needed:

- the applicant's ER

- the GEIS

- new information on thermal impacts on aquatic organisms identified by the public and other information sources

II. ACCEPTANCE CRITERIA

Acceptance criteria for the evaluation of thermal impacts on aquatic organisms are addressed in ESRP Section 4.5, Ecological Resources.

Technical Rationale

The review conducted under this ESRP leads to the preparation of SEIS sections that incorporate the conclusions in the GEIS related to impacts of thermal impacts on aquatic organisms from continued plant operations during the license renewal term, refurbishment, and alternatives to the proposed action. The review should also address any new and significant information.

III. REVIEW PROCEDURES

1. Review the discussion of thermal impacts on aquatic organisms in the GEIS. This step establishes the basis for evaluating information identified by the applicant, the public, and the staff.

2. Follow the general approach for information and analysis content for all ecology issues, described in ESRP Section 4.5, Ecological Resources, with the following additions:

- The reviewer should obtain additional information about the plant's cooling water system design and operating procedures from the applicant's environmental report and other ESRP sections that describe components of the cooling system and the hydrodynamics and physical impacts of the effluent. The following data or information may be needed to assess thermal impacts:

  - a copy of the plant's current NPDES permit

  - a copy of the plant's current 316(a) variance (if required, according to Section 316(a) of the Federal Water Pollution Control Act or CWA [33 U.S.C. 1326(a)])

  - a description of any proposed changes to the plant-cooling-system discharge

  - studies or monitoring programs of thermal effects entrainment, including information on when and where conducted and by whom, the objective (why conducted), the biological entities or attributes chosen for study, methods and results (how conducted) applicable to this license renewal application, and what data or data summaries might be available for NRC review

  - the temperature duration-mortality relationship and susceptibility of local species to heat shock

  - a description of applicable State and Federal effluent guidelines (40 CFR Part 423) and the thermal standards or limitations applicable to the water body to which the discharge is made (including maximum permissible temperature, maximum permissible temperature increase, mixing zones, and maximum rates of increase and decrease) and whether and to what extent these standards or limitations have been approved by the EPA in accordance with the CWA

- Additional information may also be needed on the specific nature of the thermal stress, such as:

  - maximum sustained temperatures for each season that are consistent with maintaining local ecological systems

  - temperature effects data for local species or their surrogates

  - thermal requirements of downstream aquatic life where upstream warming would adversely affect downstream temperature requirements

  - areal extent of the plume

  - physical factors that might concentrate biota in the plume

- Briefly summarize any site-specific thermal effluent studies, monitoring programs, or thermal effects or mortality studies and include locations, dates, objectives, methods, and results applicable to this license renewal application, and what data or data summaries might be available for NRC review. Provide estimates of numbers by taxa of fish and shellfish affected by and susceptible to the thermal effluent on a daily, monthly, and annual basis.

Provide areal or volumetric estimates of thermally affected fish and shellfish habitat. Provide full documentation of analytical or modeling techniques used to assess effects. Give perspective to these effects in terms of such concepts as the commercial, recreational, and ecosystem services they would have provided.

- If aquatic resources have been monitored, provide an analysis of temporal and geographic trends in the data that might indicate whether fish and shellfish populations have increased, decreased, or remained stable during the initial term of operation. Show any relationships between patterns of thermal effects and trends in potentially affected populations. Because entrainment, impingement, and thermal impacts all affect field populations simultaneously, provide, if possible, a single discussion of the effects of these stressors on trends in field data rather than having separate discussions for these three stressors individually.

3. Review the applicant's ER, including:

- the applicant's process for identifying new and potentially significant information

- any new information included in the ER on thermal discharge issues known to the applicant and the public

4. Prepare a statement for the SEIS that:

- describes analysis of continued plant operations and refurbishment

- summarizes the permitting documents

- states that a NPDES permit and 316(a) variance (if required) or equivalent State permits and supporting documentation are current and available

- describes measures to mitigate adverse impacts

- provides the significance level of the environmental impacts

- describes new and significant information, if any

IV. EVALUATION FINDINGS

The depth and extent of the input to the SEIS would be determined by the analysis required to reach a conclusion related to impacts of thermal impacts on aquatic organisms from continued plant operations during the license renewal term and refurbishment. The information that should be included in the SEIS is described in the review procedures.

V. IMPLEMENTATION

The method described in this ESRP would be used by the staff in evaluating conformance with the Commission's regulations, except in those cases in which the applicant for license renewal proposes an acceptable alternative for complying with specified portions of the regulations.

VI. BIBLIOGRAPHY

The bibliography for this section is provided in ESRP Section 4.5, Ecological Resources.

ENVIRONMENTAL STANDARD REVIEW PLAN

4.5.5 SPECIAL STATUS SPECIES AND HABITATS

I. AREAS OF REVIEW

This environmental standard review plan (ESRP) provides guidance for the review of impacts to threatened, endangered, and protected species (and critical habitats, if required) and essential fish habitat from continued plant operations during the renewal term and refurbishment. Impacts are discussed in the *Generic Environmental Impact Statement for License Renewal of Nuclear Plants* (GEIS; NUREG-1437, Volumes 1, 2, and 3, Revision 1).

The scope includes (1) review the discussion of threatened, endangered, and protected species and essential fish habitat in the GEIS, (2) review the applicant's environmental report (ER), (3) identify and address any new and significant information, and (4) prepare input to the supplemental environmental impact statement (SEIS).

Threatened, endangered, and protected species and essential fish habitat includes those species and habitats that are Federally protected under the Endangered Species Act (ESA) and the Magnuson-Stevens Act (MSA). These Federal acts put requirements on Federal agencies such as the NRC. State-protected species and habitats are discussed in Chapter 3 of the GEIS (and as required in a site-specific supplemental environmental impact statement [SEIS]), but not specifically in Chapter 4 because protection of these species and habitats is under the jurisdiction of the State agencies, which do not put requirements on NRC. Chapter 3 of the SEIS should describe the threatened, endangered, and protected species and essential fish habitat while this section of the SEIS should discuss the corresponding environmental consequences and impacts.

The ESA is administered by the U.S. Fish and Wildlife Service (USFWS) and the National Marine Fisheries Service (NMFS), referred to collectively as the "Services." Section 7(a)(2) of the ESA states that each Federal agency shall, in consultation with the Secretary (Secretary of the Interior or Secretary of Commerce), insure that any action they authorize, fund, or carry out is not likely to jeopardize the continued existence of a listed species or result in the destruction or adverse modification of designated critical habitat. Marine and anadromous species are generally the responsibility of NMFS, and all other species and habitats the responsibility of USFWS. While the analyses of environmental impact under NEPA and NRC's regulations in the SEIS would include an assignment of NRC impact levels of SMALL, MODERATE, or LARGE for each issue, the assessment of endangered species under ESA must include a determination of "jeopardy" or "no jeopardy" for the species itself and "adverse modification" or "no adverse modification" for designated critical habitat based on a careful analysis of the best available scientific and commercial data. Guidance for conducting consultations with the Services can be found in *The Endangered Species Consultation Handbook* (USFWS 1998). Note that an SEIS for license renewal can have separate consultations with the two Services for different species.

The MSA mandates emphasize wetlands, anadromous fish habitat, and habitat of other Federally-managed marine and estuarine species. The essential fish habitat (EFH) consultation process is very similar to the Section 7 consultation process for endangered species. Guidance for conducting EFH consultations with NMFS can be found in *Essential Fish Habitat Consultation Guidance, Version 1.1* (NMFS 2004). Similar to ESA, while the analyses of environmental impact under National Environmental Policy Act (NEPA) and NRC's regulations in the SEIS would include an assignment of NRC impact levels of SMALL, MODERATE, or LARGE for each issue, the EFH assessment under MSA must include a determination of either "adverse impact" or "no adverse impact" to EFH. Adverse effect is any impact that reduces the quality and/or quantity of essential fish habitat. Adverse effects may include direct or indirect physical, chemical, or biological alterations of the waters or substrate and loss of, or injury to, benthic organisms, prey species and their habitat, and other ecosystem components, if such modifications reduce the quality and/or quantity of EFH. Withdrawal of water for power plant cooling, even with return after passing through the cooling water system, is generally considered to decrease the quantity of both fish habitat and prey species. A finding of adverse impact should be qualified in terms of its magnitude relative to the stock or population of interest as "minimal," "less than substantial," or "substantial."

Designated EFH may overlap with the habitat (including critical habitat) of species listed as threatened or endangered under the ESA. In such cases, an assessment can be prepared that integrates the biological assessment under ESA and the EFH assessment under MSA. In addition, an integrated consultation process can be conducted with National Oceanic and Atmospheric Administration.

Special status species and habitats are reported in three parts of the SEIS. They should be described as part of the affected environment in Chapter 3. This description can include life history summaries in which the level of detail is appropriate for the level of analysis that would follow. They should also be included among the ecological entities that are assessed in Chapter 4. Communications with the Services regarding either listed species or EFH should be included in an appendix to the SEIS. If the geographic range of either species listed for Federal protection or essential fish habitat include the plant site, then the biological assessment and EFH assessment should also be included in that appendix. Compared to the information in the main body of the SEIS, the assessments should focus sharply on the species or habitat and relevant information needed to support the assessment.

NRC conducts ESA and MSA consultations with USFWS and NMFS. Endangered or threatened species, critical habitats, and essential fish habitat should all be assessed in the SEIS. In addition, if endangered or threatened species or their critical habitats occur in the vicinity of the nuclear plant site, a biological assessment should be prepared and included as an appendix in the SEIS. Similarly, if EFH occurs in the vicinity of the nuclear plant site, an EFH assessment should be prepared and included as an appendix in the SEIS.

Data and Information Needs

The types of data and information needed would be affected by nuclear power plant site- and plant-specific factors. The following data or information may be needed:

- the applicant's ER

- the GEIS

- new information on impacts on threatened, endangered, and protected species and essential fish habitat identified by the public and other information sources

II. ACCEPTANCE CRITERIA

Acceptance criteria for the evaluation of threatened, endangered, and protected species and essential fish habitat are addressed in ESRP Section 4.5, Ecological Resources.

Technical Rationale

The review conducted under this ESRP leads to the preparation of SEIS sections that incorporate the conclusions in the GEIS related to impacts on threatened, endangered, and protected species and essential fish habitat from continued plant operations during the license renewal term, refurbishment, and alternatives to the proposed action. The review should also address any new and significant information.

III. REVIEW PROCEDURES

Suggested steps for the ESA and MSA review process are as follows:

IIIa. Review Procedures for ESA

1. Review the discussion of threatened, endangered, and protected species in the GEIS. This step establishes the basis for evaluating information identified by the applicant, the public, and the staff.

2. Follow the general approach for information and analysis content for all ecology issues, described in ESRP Section 4.5, Ecological Resources, with the following additions:

- Identify potential environmental effects from continued operations during the renewal term that might impact threatened or endangered species or critical habitat.

- Initiate informal consultation:

    - Initiate informal consultation with the USFWS and, as applicable, NMFS to determine the likelihood of adverse effects on listed, candidate, and proposed threatened or endangered species or critical habitats. The NRC would provide the USFWS and NMFS with a description of nuclear plant operations during the renewal term, a description of any listed, candidate, and proposed threatened or endangered species or critical habitats that it knows to exist on or in the vicinity of the power plant site, and a description of the manner in which operations during the renewal term is anticipated to impact listed, candidate, and proposed threatened or endangered species or critical habitats.

      A detailed description of the procedural requirements for consultation under the ESA, including definitions of listed, candidate, and proposed threatened and endangered species and critical habitats, jeopardy, adverse modification, etc., are found in *Endangered Species Act Consultation Handbook: Procedures for Conducting Section 7 Consultations and Conferences* (USFWS 1998). This document also contains guidelines for procuring an incidental take permit.

    - Initiate a biological assessment if one is requested by the USFWS or NMFS as a prerequisite to making a finding in the informal consultation. A biological assessment is required if listed threatened or endangered species or critical habitats are present in environment affected by the proposed action. It is optional only if proposed threatened or endangered species or critical habitats are involved.

      The biological assessment should address listed and proposed threatened or endangered species and critical habitats that are likely to be affected. The content of biological assessments prepared pursuant to the ESA is at the discretion of the NRC, although recommended contents may be found in 50 CFR 402.12, "Interagency Cooperation—Endangered Species Act of 1973, as amended; Biological Assessments."

- Initiate formal consultation

    - Initiate formal consultation with the USFWS and NMFS, as applicable, to determine if operations during the renewal term are likely to jeopardize the continued existence of a listed species (jeopardy) or destroy or adversely modify critical habitat (adverse modification). The USFWS and/or NMFS would render a biological opinion of jeopardy/no jeopardy and adverse modification/no adverse modification and issue an incidental take statement as warranted. The USFWS and/or NMFS would also provide a list of reasonable and prudent alternatives (e.g., alternative design and/or placement of structures, alternative schedules, or alternative operational procedures, habitat improvement, etc.) for avoiding jeopardy or adverse modification, and minimizing incidental take. Alternatively, the USFWS and/or NMFS may provide a statement that no reasonable and prudent alternatives exist, along with an explanation.

If the USFWS and/or NMFS renders an opinion of no jeopardy and no adverse modification, then prepare a statement for the SEIS that (1) summarizes the information that has been reviewed, the analyses that have been conducted, and the basis for the decision rendered by the USFWS and/or NMFS, and (2) concludes that operations during the renewal term would not jeopardize the continued existence of a listed species or destroy or adversely modify critical habitat.

If the USFWS and/or NMFS renders an opinion of jeopardy and/or adverse modification, prepare a statement for the SEIS that (1) summarizes the information that has been reviewed, the analyses that have been conducted, and the basis for the decision rendered by the USFWS and/or NMFS, (2) includes a list of reasonable and prudent alternatives for avoiding jeopardy and/or adverse modification, or provides a statement that no reasonable and prudent alternatives exist, (3) includes a statement regarding the magnitude of anticipated incidental take and a list of reasonable and prudent alternatives for avoiding or reducing incidental take, and, if needed, procurement of an incidental take permit under sections 10(a)(1)(A) and 10(a)(1)(B) of the ESA, and (4) concludes that operations during the renewal term would jeopardize the continued existence of a listed species and/or destroy or adversely modify critical habitat.

3. Review the applicant's ER, including:

    - the applicant's process for identifying new and potentially significant information

    - any new information included in the ER on threatened, endangered, and protected species and essential fish habitat issues known to the applicant and the public

4. Prepare a statement for the SEIS that:

    - describes analysis of continued plant operations and refurbishment and summarizes the information that has been reviewed, the analyses that have been conducted, and the basis for the opinion rendered by the USFWS. This section of the SEIS should present:

        - a list of adverse impacts to listed, candidate, and proposed threatened or endangered species or critical habitats from continued operations during the renewal term and refurbishment

        - a list of the impacts for which there are reasonable and prudent alternatives to limit adverse effects and the associated alternatives

        - the applicant's commitments to limit these impacts

    - describes measures to mitigate adverse impacts

    - provides the significance level of the environmental impacts

    - describes new and significant information, if any

IIIb. Review Procedures for MSA

Procedures for EFH assessments are very similar to those for threatened and endangered species and can often be performed simultaneously. Suggested steps for the review process are as follows:

1. Review the discussion of essential fish habitat in the GEIS. This step establishes the basis for evaluating information identified by the applicant, the public, and the staff. They should include results of requests to the NMFS for EFH in the project area.

2. Follow the general approach for information and analysis content for all ecology issues, described in ESRP Section 4.5, Ecological Resources, with the following additions:

   • Identify potential environmental effects related to operations during the renewal term that might adversely affect EFH. Describe EFH that might be affected by continued plant operation. Include EFH and the species for which it is designated among the biological entities to be analyzed for each aquatic issue. EFH regulations "further clarify EFH by defining 'waters' to include aquatic areas and their associated physical, chemical, and biological properties that are used by fish and may include aquatic areas historically used by fish where appropriate; 'substrate' to include sediment, hard bottom, structures underlying the waters, and associated biological communities; 'necessary' to mean the habitat required to support a sustainable fishery and the managed species' contribution to a healthy ecosystem; and 'spawning, breeding, feeding, or growth to maturity' to cover a species' full life cycle" (NMFS 2004).

   • Initiate abbreviated consultation with the NMFS:

     - Initiate abbreviated consultation with the NMFS to determine the likelihood of adverse effects on EFH. The NRC would provide NMFS with a description of operations during the renewal term, a description of any EFH or presence of EFH species that it knows to exist on or in the vicinity of the site, and a description of the manner in which operations during the renewal term is anticipated to impact EFH. A detailed description of the procedural requirements for consultation, including definitions, is found in *Essential Fish Habitat Consultation Guidance, Version 1.1* (NMFS 2004).

     - Initiate an EFH assessment, if one is requested by the NMFS as a prerequisite to making a finding in the informal consultation. An EFH assessment is required if EFH may be present in the area to be affected by the action.

       The assessment should include: the results of an onsite inspection; the views of recognized experts on the habitat or species effects; a literature review; an analysis of alternatives to the proposed action; and any other relevant information.

       EFH assessments must include:

       - a description of the proposed action

- - an analysis of the effects, including cumulative effects, of the action on EFH, the managed species, and associated species by life history stage

  - the NRC's views regarding the effects of the action on EFH

  - proposed mitigation, if applicable

- Initiate expanded consultation with the NMFS

  - Initiate expanded consultation with the NMFS to determine if operations during the renewal term are likely to adversely affect EFH. The Service would render a written finding of adverse effects or no adverse effects and Conservation Recommendations as warranted. The MSA mandates that a federal action agency (NRC) must respond to NMFS's proposed EFH Conservation Recommendations in writing within 30 days.

    NRC's response must either include a description of measures proposed by the agency for avoiding, mitigating, or offsetting the impact of the activity on EFH or else explain its reasons for not following the recommendations, including the scientific rationale for any disagreements with NMFS over the anticipated effects of the proposed action and the measures needed to offset such effects. If an agency decision is inconsistent with a NMFS Conservation Recommendation, the NMFS Director may request a meeting with the Commission to further discuss the project.

    EFH Assessments must include references to:

    - an EFH that may be found in water bodies that may be affected by plant operation

    - license renewal activities and modifications to plant operation that may adversely affect EFH

    - letters and communications with the NMFS and any resulting NMFS memoranda (include any letters in the appendix to the ER)

3. Review the applicant's ER, including:

   - the applicant's process for identifying new and potentially significant information

   - any new information included in the ER on impingement, entrainment, and thermal discharge issues known to the applicant and the public

4. Prepare a statement for the SEIS that:

   - describes analysis of continued plant operations and refurbishment and summarizes the information that has been reviewed, the analyses that have been conducted, and the basis for the opinion rendered by the NMFS. This section of the SEIS should present:

     - a list of adverse impacts to EFH from continued operations during the renewal term and refurbishment

     - a list of the impacts for which there are Conservation Recommendations to limit adverse effects

- the applicant's commitments to limit these impacts
- describes measures to mitigate adverse impacts
- provides the significance level of the environmental impacts
- describes new and significant information, if any

IV. EVALUATION FINDINGS

The depth and extent of the input to the SEIS would be determined by the analysis required to reach a conclusion related to impacts of continued plant operations during the license renewal term and refurbishment. The information that should be included in the SEIS is described in the review procedures.

V. IMPLEMENTATION

The method described in this ESRP would be used to evaluate conformance with the Commission's regulations, except in those cases in which the applicant for license renewal proposes an acceptable alternative for complying with specified portions of the regulations.

VI. BIBLIOGRAPHY

50 CFR Part 402. *Code of Federal Regulations*, Title 50, *Wildlife and Fisheries*, Part 402, "Interagency Cooperation Endangered Species Act of 1973, as Amended."

Coastal Zone Management Act. 16 USC 1451 et seq.

Council on Environmental Quality (CEQ). 1993. *Incorporating Biodiversity Considerations into Environmental Impact Analysis under the National Environmental Quality Act.* CEQ, Executive Office of the President, Washington, D.C.

Endangered Species Act. 16 USC 1531 et seq.

Federal Water Pollution Control Act (FWPCA) (also known as Clean Water Act). 33 USC 1251 et seq.

Magnuson-Stevens Fishery Conservation and Management Act. 16 USC 1801 et seq.

Menzie, C.A., et al. (10 other authors). 1996. "Special Report of the Massachusetts Weight-of-Evidence Workgroup: A Weight of Evidence Approach for Evaluating Ecological Risks." *Human and Ecological Risk Assessment* 2(2):277–304.

National Environmental Policy Act of 1969. 42 U.S.C. 4321 et seq.

National Marine Fisheries Service (NMFS). 2004. *Essential Fish Habitat Consultation Guidance, Version 1.1*. Office of Habitat Conservation, Silver Spring, Maryland.

Suter, G.W. 1993. Chapter 10: Retrospective Risk Assessment. Pp. 311–364. In G.W. Suter, II (ed.). *Ecological Risk Assessment*. Lewis Publishers, Boca Raton, Florida.

U.S. Environmental Protection Agency (EPA). 1973. *Water Quality Criteria, 1972*. EPA-R3-73-033, Ecological Research Series. Committee on Water Quality Criteria, National Academy of Sciences and National Academy of Engineering, Washington, D.C.

U.S. Environmental Protection Agency (EPA). 1998. *Guidelines for Ecological Risk Assessment*. EPA/630/R-95/002. Risk Assessment Forum, Washington, D.C.

U.S. Fish and Wildlife Service (USFWS). 1998. *Endangered Species Act Consultation Handbook (Washington, D.C.): Procedures for Conducting Section 7 Consultations and Conferences* (Final).

U.S. NUCLEAR REGULATORY COMMISSION
ENVIRONMENTAL STANDARD
REVIEW PLAN
OFFICE OF NUCLEAR REACTOR REGULATION

4.6 HISTORIC AND CULTURAL RESOURCES

I. AREAS OF REVIEW

This environmental standard review plan (ESRP) provides guidance for the review of potential impacts on historic and cultural resources during the renewal term and refurbishment. Impacts are discussed in the *Generic Environmental Impact Statement for License Renewal of Nuclear Plants* (GEIS; NUREG-1437, Volumes 1, 2, and 3, Revision 1).

The scope includes (1) review of the discussion of historic and cultural resources in the GEIS, (2) review of the applicant's environmental report (ER), (3) identifying and addressing any new and significant information, and (4) preparing input to the supplemental environmental impact statement (SEIS).

Section 106 of the National Historic Preservation Act of 1966 (NHPA), as amended (16 U.S.C. 470 et seq.), requires that Federal agencies take into account the effects of the agency's undertaking on properties included in or eligible for the *National Register of Historic Places* (NRHP) and, prior to approval of an undertaking, to afford the Advisory Council on Historic Preservation (ACHP) a reasonable opportunity to comment on the undertaking. The issuance of a renewed operating license for a nuclear power plant is an undertaking that could possibly affect either known or currently undiscovered historic properties.

An assessment of impact for license renewal differs from that of original licensing. In many cases, nuclear power plant sites were not investigated prior to plant construction. Even though the NHPA became law in 1966, the enacting regulations located at 36 CFR Part 800 were not available until 1979. By this point, many of the operating reactors today were constructed and operating. Therefore, most applicants are not aware of the occurrence or status of historic and cultural resources on their site.

In accordance with 36 CFR 800.8(c) "Use of the NEPA process for section 106 purposes," the NRC coordinates its Section 106 responsibilities under the National Environmental Policy Act (NEPA) for license renewal reviews. The NRC may use the NEPA process to comply with Section 106 in lieu of the procedures set forth in Sections 800.3 through 800.6 provided all consulting parties (ACHP, State Historic Preservation Officer [SHPO], Tribal Historic Preservation Officer [THPO], American Indian Tribes, the public, and other interested stakeholders) have been notified in advance and it meets the standards of 36 CFR 800.8(c). The NRC will consult with the appropriate SHPO/THPO for each plant-specific license renewal review. Through early coordination all issues will be identified.

The area(s) within which historic and cultural resources should be identified is referred to as the Area of Potential Effect(s) (APE), defined at 36 CFR 800.16(d) as the geographic area or areas within which an undertaking may directly or indirectly cause alterations in the character or use of important cultural resources, if any such resources exist. The APE is influenced by the scale and nature of an undertaking and may be different for different kinds of effects caused by the undertaking (36 CFR 800.16(d)). For NRC reviews, the APE is defined as the area at the power plant site and its immediate environs that may be impacted by post-license renewal land-disturbing operations or projected refurbishment activities. The APE may extend beyond the immediate environs in those instances where post- license-renewal land-disturbing operations or projected refurbishment activities specifically related to license renewal may potentially have an effect on known or proposed historic sites. This determination is made irrespective of ownership or control of the lands of interest.

The purpose of the historic and cultural resources assessment is to ensure that such resources that are considered eligible for inclusion in the NRHP are not adversely affected by proposed activities related to renewal term operations. If adverse effects cannot be avoided, mitigation must be developed in consultation with the appropriate SHPO/THPO and other interested parties. Historic and cultural resources may include prehistoric or historic archaeological sites, historic properties, districts, and landscapes, as well as traditional cultural properties that may have significance for American Indian Tribes or Native Hawaiian organizations.

Data and Information Needs

The types of data and information needed would be affected by nuclear power plant site- and plant-specific factors. The following data or information may be needed:

- the applicant's ER

- the GEIS

- new information identified by the public and other information sources

- a map that identifies the APE and a site disturbance map

- applicant's cultural resource protection procedures or Cultural Resource Management Plans

- previous cultural resources investigations that have identified historic and cultural resources, along with site-specific locations for those resources that are either located in or near the APE

- information related to past evaluations for eligibility for the NRHP (36 CFR 60), and associated consultations with the SHPO, local preservation officials, or American Indian Tribal officials

II. ACCEPTANCE CRITERIA

Acceptance criteria for the review of historic and cultural resources impacts during the renewal term are based on the relevant requirements of the following:

- 10 CFR 51.45(c), "Analysis." The environmental report must include an analysis that considers and balances the environmental effects of the proposed action, the environmental impacts of alternatives to the proposed action, and alternatives available for reducing or avoiding adverse environmental effects.

- 10 CFR 51.53(c)(2). The report must contain a description of the proposed action, including the applicant's plans to modify the facility or its administrative control procedures as described in accordance with 10 CFR 54.21 of this chapter. This report must describe in detail the affected environment around the plant, the modifications directly affecting the environment or any plant effluents, and any planned refurbishment activities. In addition, the applicant shall discuss in this report the environmental impacts of alternatives and any other matters discussed in 10 CFR 51.45.

- 10 CFR 51.53(c)(3)(ii)(K). All applicants shall identify any potentially affected historic or archaeological properties and assess whether any of these properties will be affected by future plant operations and any planned refurbishment activities in accordance with the National Historic Preservation Act.

- 10 CFR 51.70(b). The draft environmental impact statement will be concise, clear, and analytic, and written in plain language with appropriate graphics....The format provided in Section 1(a) of Appendix A of this subpart should be used. The NRC staff will independently evaluate and be responsible for the reliability of all information used in the draft environmental impact statement.

- 10 CFR 51.71(d), concerning the draft environmental impact statement, will include a preliminary analysis that considers and weighs the environmental effects of the proposed action, the environmental impacts of alternatives to the proposed action, and alternatives available for reducing or avoiding adverse environmental effects.

- 10 CFR 51.95(c), concerning renewal of an operating license or combined license for a nuclear power plant. Under Parts 52 or 54 of this chapter, the Commission shall prepare an environmental impact statement, which is a supplement to the Commission's NUREG–1437, "Generic Environmental Impact Statement for License Renewal of Nuclear Plants."

- 10 CFR Part 51, Appendix A to Subpart A, para. 7, concerning the environmental consequences of alternatives, including the proposed actions and any mitigating actions which may be taken. Alternatives eliminated from detailed study will be identified and a discussion of those alternatives will be confined to a brief statement of the reasons why the alternatives were eliminated. The level of information for each alternative considered in detail will reflect the depth of analysis required for sound decisionmaking.

- 10 CFR Part 51, Appendix B to Subpart A, "Environmental Effect of Renewing the Operating License of a Nuclear Power Plant," Table B-1, "Summary of Findings on NEPA Issues for License Renewal of Nuclear Power Plants"

- National Historic Preservation Act of 1966 (16 U.S.C. 470 et seq.).

- 36 CFR Part 800, "Protection of Historic Properties," specifically 36 CFR 800.8(c), "Use of the NEPA Process for Section 106 Purposes"

- 36 CFR Part 63, containing guidance by which historic properties are evaluated and determined eligible for listing on the NRHP

Technical Rationale

The review conducted under this ESRP leads to the preparation of SEIS sections that incorporate the conclusions in the GEIS related to impacts on historic and cultural resources from continued plant operations during the license renewal term, refurbishment, and alternatives to the proposed action. The review should also address any new and significant information.

III. REVIEW PROCEDURES

To analyze the impact of plant operations during the renewal term on historic and cultural resources, review the information collected and discussed in Section 3.7 of the ESRP and complete the following steps:

1. Review the discussion of the impacts of plant operations during the renewal term on historic and cultural resources in the GEIS to identify the information considered and the conclusions reached. This step establishes the base for evaluation of information identified by the applicant, the public, and the staff. Category 2 historic and cultural resource issues identified in the GEIS are listed in the following table.

Table B-1. Summary of Findings on NEPA Issues for License Renewal of Nuclear Power Plants		
Issue	**Category**	**Finding**
Historic and Cultural Resources		
Historic and cultural resources	2	Continued operations and refurbishment associated with license renewal are expected to have no more than small impacts on historic and cultural resources located onsite and in the transmission line ROW because most impacts could be mitigated by avoiding those resources. The National Historic Preservation Act (NHPA) requires the Federal agency to consult with the State Historic Preservation Officer (SHPO) and appropriate Native American Tribes to determine the potential effects on historic properties and mitigation, if necessary.

2. Analyze the historic and cultural resources impacts associated with continued plant operations during the renewal term and refurbishment, as follows:

   - Determine the direct and indirect impacts that could result from continued plant operations and refurbishment activities associated with license renewal, e.g., building new waste storage facilities, new parking areas, new access roads to existing transmission lines, or new transmission lines.

   - Review any issues related to historic and cultural resources identified during the public scoping period.

   - Identify historic and cultural resources through consultation with the SHPO/THPO and by reviewing any archaeological investigations or building surveys conducted within the APE (power plant site) and the immediate environs.

   - Review the site disturbance map (developed by a qualified archaeologist) that indicates areas of heavy disturbance and areas of high potential for undiscovered historic and cultural resources.

   - Review any correspondence from the SHPO/THPO, any American Indian Tribes, or local preservation officials regarding any archaeological investigations or building surveys conducted on the applicant's site.

   - If significant resources are located within the APE, review any procedures or integrated cultural resources management plans instituted by the applicant to protect the historic and cultural resources identified on the site or within the in-scope transmission line right-of-way. Also, verify that the applicant has developed these procedures and plans in consultation with the appropriate SHPO, local preservation official, or American Indian Tribes.

   - Evaluate the historic significance of historic and cultural resources by applying the NRHP criteria. Also, apply criteria for historic and cultural resources within the APE that have not been previously evaluated for NRHP eligibility.

   - Through consultation with American Indian Tribes, identify any traditional cultural properties.

3. Evaluate the effects of continued operations and refurbishment to historic and cultural resources. If historic and cultural resources are found to be in or near APE, the assessment of effects can be conducted using criteria for effect and adverse effect contained in 36 CFR 800.5, "Assessment of Adverse Effects." Assessments of effects should involve consultation with the SHPO/THPO, local historic preservation officials, and American Indian Tribal members, as necessary, to ensure that all potential values are identified for specific resources. The assessment would result in one of the following conclusions:

   - No effect; the proposed license-renewal-term operations activities would not affect any known significant historic and cultural resources.

   - No adverse effect; the activities would affect one or more historic or cultural resources, but the effect would not significantly alter the historic character of the resource(s).

- Adverse effect; the proposed activities would result in harm to the qualities that make one or more historic or cultural resource significant.

4. If an adverse effect would result from continued plant operations during the license renewal term and refurbishment, the applicant, in consultation with the NRC, SHPO, American Indian Tribes, and other interested parties should identify strategies to avoid, minimize or mitigate the impacts to significant historic and cultural resources.

5. Review the applicant's ER, including:

   - the applicant's process for identifying new and potentially significant information

   - any new information included in the ER on historic and cultural issues known to the applicant, the public, and other consulting parties

   - all correspondence initiated by the applicant to the SHPO, American Indian Tribes, or local preservation officials

   - a map that identifies the APE and power plant property boundary

   - a map that documents the level of past ground disturbance in the APE

   - cultural resource protection procedures or Cultural Resource Management Plans

   - previous cultural resources investigations that have identified historic and cultural resources, along with site-specific locations for those resources that are either located in or near the APE information related to past evaluations for eligibility for the NRHP (36 CFR 60), and associated consultations with the SHPO, local preservation officials, or American Indian Tribal officials

6. Prepare a statement for the SEIS that:

   - describes analysis of continued plant operations and refurbishment and summarizes the information that has been reviewed, and the analyses that have been conducted

   - describes measures to mitigate adverse impacts

   - provides the significance level of the environmental impacts

   - discusses new and significant information, if any

IV. EVALUATION FINDINGS

The depth and extent of the information in the assessment would be governed by the extent and significance of the effects of operations during the renewal term on historic and cultural resources. The reviewer should verify that sufficient information is available to meet the relevant requirements and that the SEIS includes the information described under the review procedures.

V. IMPLEMENTATION

The method described in this ESRP would be used by the staff in evaluating conformance with the Commission's regulations, except in those cases in which the applicant for license renewal proposes an acceptable alternative for complying with specified portions of the regulations.

VI. BIBLIOGRAPHY

10 CFR Part 51. *Code of Federal Regulations*, Title 10, *Energy,* Part 51, "Environmental Protection Regulations for Domestic Licensing and Related Regulatory Functions."

36 CFR Part 60. *Code of Federal Regulations*, Title 36, *Parks, Forests, and Public Properties*, Part 60, "National Register of Historic Places."

36 CFR Part 63. *Code of Federal Regulations*, Title 36, *Parks, Forests, and Public Properties*, Part 63, "Determinations of Eligibility for Inclusion in the National Register of Historic Places."

36 CFR Part 800. *Code of Federal Regulations*, Title 36, *Parks, Forests, and Public Properties*, Part 800, "Protection of Historic Properties."

National Environmental Policy Act of 1969 (NEPA). 42 USC 4321 et seq.

National Historic Preservation Act of 1966. (16 USC 470 et seq.).

U.S. Nuclear Regulatory Commission (NRC). 2013. *Generic Environmental Impact Statement for License Renewal of Nuclear Plants.* NUREG-1437, Vols. 1, 2, and 3, Revision 1. Office of Nuclear Reactor Regulation, Washington, D.C.

U.S. NUCLEAR REGULATORY COMMISSION
ENVIRONMENTAL STANDARD REVIEW PLAN
OFFICE OF NUCLEAR REACTOR REGULATION

4.7 SOCIOECONOMICS

I. AREAS OF REVIEW

This environmental standard review plan (ESRP) guides the review of socioeconomic impacts of continued plant operations and refurbishment associated with license renewal. Socioeconomic impacts of continued plant operations during the license renewal term and refurbishment are evaluated in the *Generic Environmental Impact Statement for License Renewal of Nuclear Plants* (GEIS; NUREG-1437, Volumes 1, 2, and 3, Revision 1).

The scope includes: (1) review of potential socioeconomic impacts of continued plant operations and refurbishment addressed in the GEIS, (2) evaluation of new information for significance, and (3) preparation of input to the supplemental environmental impact statement (SEIS).

Data and Information Needs

The types of data and information needed would be affected by nuclear power plant site- and plant-specific factors. The following data or information may be needed:

- the applicant's ER

- the GEIS

- new information on socioeconomic factors identified by the public and other information sources

I. ACCEPTANCE CRITERIA

Acceptance criteria for the evaluation of the socioeconomic impacts of continued nuclear power plant operations during the operating license renewal term and refurbishment are based on the following regulations:

- 10 CFR 51.45(c), "Analysis." The environmental report must include an analysis that considers and balances the environmental effects of the proposed action, the environmental impacts of alternatives to the proposed action, and alternatives available for reducing or avoiding adverse environmental effects.

- 10 CFR 51.53(c)(2). The report must contain a description of the proposed action, including the applicant's plans to modify the facility or its administrative control procedures as described in accordance with 10 CFR54.21 of this chapter. This report must describe in detail the affected environment around the plant, the modifications directly affecting the environment or any plant effluents, and any planned refurbishment activities. In addition, the applicant shall discuss in this report the environmental impacts of alternatives and any other matters discussed in 10 CFR 51.45.

- 10 CFR 51.70(b). The draft environmental impact statement will be concise, clear, and analytic, and written in plain language with appropriate graphics.…The format provided in Section 1(a) of Appendix A of this subpart should be used. The Nuclear Regulatory Commission (NRC) staff will independently evaluate and be responsible for the reliability of all information used in the draft environmental impact statement.

- 10 CFR 51.71(d). The draft environmental impact statement will include a preliminary analysis that considers and weighs the environmental effects of the proposed action; the environmental impacts of alternatives to the proposed action; and alternatives available for reducing or avoiding adverse environmental effects.

- 10 CFR 51.95(c), concerning renewal of an operating license or combined license for a nuclear power plant. Under Parts 52 or 54 of this chapter, the Commission shall prepare an environmental impact statement, which is a supplement to the Commission's NUREG-1437, "Generic Environmental Impact Statement for License Renewal of Nuclear Plants."

- 10 CFR Part 51, Appendix A to Subpart A, para. 7, concerning the environmental consequences of alternatives, including the proposed actions and any mitigating actions which may be taken. Alternatives eliminated from detailed study will be identified and a discussion of those alternatives will be confined to a brief statement of the reasons why the alternatives were eliminated. The level of information for each alternative considered in detail will reflect the depth of analysis required for sound decisionmaking.

- 10 CFR Part 51, Appendix B to Subpart A, "Environmental Effect of Renewing the Operating License of a Nuclear Power Plant," Table B-1, "Summary of Findings on NEPA Issues for License Renewal of Nuclear Power Plants"

<u>Technical Rationale</u>

The review conducted under this ESRP leads to the preparation of SEIS sections that incorporate the conclusions in the GEIS related to socioeconomic impacts from continued plant operations during the license renewal term, refurbishment, and alternatives to the proposed action. The review should also address any new and significant information.

III. REVIEW PROCEDURES

Suggested steps for the socioeconomic review process are as follows:

1. Review the discussion of potential socioeconomic impacts of continued plant operations during the operating license renewal term in the GEIS. This step establishes the basis for evaluating any new and significant socioeconomic information identified by the applicant, the public, and the staff. Category 1 socioeconomic issues identified in the GEIS are listed in the following table.

Table B-1. Summary of Findings on NEPA Issues for License Renewal of Nuclear Power Plants		
Issue	**Category**	**Finding**
Socioeconomics		
Employment and income, recreation and tourism	1	SMALL. Although most nuclear power plants have large numbers of employees with higher than average wages and salaries, impacts to employment, income, recreation, and tourism from continued operations and refurbishment associated with license renewal are expected to be small.
Tax revenues	1	SMALL. Nuclear plants provide tax revenue to local jurisdictions in the form of property tax payments, payments in lieu of tax (PILOT), or tax payments on energy production. The amount of tax revenue paid during the license renewal term as a result of continued operations and refurbishment associated with license renewal is not expected to change.
Community services and education	1	SMALL. Changes resulting from continued operations and refurbishment associated with license renewal to local community and educational services would be small. With little or no change in employment at the licensee's plant, value of the power plant, payments on energy production, and PILOT payments expected during the license renewal term, community and educational services would not be affected by continued power plant operations.
Population and housing	1	SMALL. Changes resulting from continued operations and refurbishment associated with license renewal to regional population and housing availability and value would be small. With little or no change in employment at the licensee's plant expected during the license renewal term, population and housing availability and values would not be affected by continued power plant operations.
Transportation	1	SMALL. Changes resulting from continued operations and refurbishment associated with license renewal to traffic volumes would be small.

2. Determine whether there is any new socioeconomic information that should be evaluated. The following sources of information should be included in the search for new information:

- The applicant's ER. An applicant is required by 10 CFR 51.53(c)(3)(iv) to disclose new and significant information regarding the environmental impacts of license renewal of which it is aware. In reviewing the applicant's ER, consider the applicant's process for discovering new information and evaluating the significance of any new information. Is the process adequate to ensure a reasonable likelihood that the applicant would be aware of new information?

- Records of public scoping meetings and correspondence related to the operating license renewal application. Compare the socioeconomic information presented by the public with information considered in the GEIS. Does the information postdate the analysis leading to the GEIS?

- Environmental quality standards and regulations. Have the applicable environmental quality standards and regulations changed since the analysis leading to the GEIS? If so, would these changes affect the NRC license renewal environmental review?

3. Evaluate the significance of new socioeconomic information.

4. Review the applicant's ER, including:

- the applicant's process for identifying new and potentially significant information

- any new information included in the ER on socioeconomic issues known to the applicant and the public

5. Prepare a statement for the SEIS that:

- describes the analysis of continued plant operations and refurbishment and summarizes the information that has been reviewed, and the analyses that may have been conducted

- describes measures to mitigate any adverse impacts

- provides the significance level of environmental impacts

- describes new and significant information, if any

IV. EVALUATION FINDINGS

The amount of information in the SEIS would be governed by the extent of the analysis required to reach a conclusion related to the potential impacts of continued plant operations on socioeconomic conditions in the region around the nuclear plant site during the license renewal term and refurbishment. The information that should be included in the SEIS is described in the review procedures.

V. IMPLEMENTATION

The method described in this ESRP would be used to evaluate conformance with the Commission's regulations, except in those cases in which the applicant for license renewal proposes an acceptable alternative for complying with specified portions of the regulations.

VI. BIBLIOGRAPHY

10 CFR Part 51. *Code of Federal Regulations*, Title 10, *Energy,* Part 51, "Environmental Protection Regulations for Domestic Licensing and Related Regulatory Functions."

National Environmental Policy Act of 1969 (NEPA). 42 USC 4321 et seq.

U.S. Nuclear Regulatory Commission (NRC). 2013. *Generic Environmental Impact Statement* (GEIS) *for License Renewal of Nuclear Plants*. NUREG-1437, Vols. 1, 2, and 3, Revision 1. Office of Nuclear Reactor Regulation, Washington, D.C.

ENVIRONMENTAL STANDARD REVIEW PLAN

OFFICE OF NUCLEAR REACTOR REGULATION

4.8 HUMAN HEALTH

I. AREAS OF REVIEW

This environmental standard review plan (ESRP) provides guidance for the analysis and assessment of the human-health impacts from continued plant operations during the license renewal term and refurbishment. Human health impacts are evaluated in the *Generic Environmental Impact Statement for License Renewal of Nuclear Plants* (GEIS; NUREG-1437, Volumes 1, 2, and 3, Revision 1).

The scope includes: (1) review of human health impacts from continued plant operations during the license renewal term and refurbishment in the GEIS, (2) evaluation of new information for significance, and (3) preparation of input to the supplemental environmental impact statement (SEIS).

Data and Information Needs

The types of data and information needed would be affected by nuclear power plant site- and plant-specific factors. The following data or information may be needed:

- the applicant's environmental report (ER)

- the GEIS

- new information on human health impacts identified by the public and other information sources

II. ACCEPTANCE CRITERIA

Acceptance criteria for the evaluation of human-health impacts are based on the relevant requirements of the following regulations:

- 10 CFR 51.45(c), "Analysis." The environmental report must include an analysis that considers and balances the environmental effects of the proposed action, the environmental impacts of alternatives to the proposed action, and alternatives available for reducing or avoiding adverse environmental effects.

- 10 CFR 51.53(c)(2). The report must contain a description of the proposed action, including the applicant's plans to modify the facility or its administrative control procedures as described in accordance with 10 CFR 54.21 of this chapter. This report must describe in detail the affected environment around the plant, the modifications directly affecting the environment or any plant effluents, and any planned refurbishment activities. In addition, the applicant shall discuss in this report the environmental impacts of alternatives and any other matters discussed in 10 CFR 51.45.

- 10 CFR 51.53(c)(3)(ii)(G). If the applicant's plant uses a cooling pond, lake, or canal or discharges into a river, an assessment of the impact of the proposed action on public health from thermophilic organisms in the affected water must be provided.

- 10 CFR 51.53(c)(3)(ii)(H). If the applicant's transmission lines that were constructed for the specific purpose of connecting the plant to the transmission system do not meet the recommendations of the National Electric Safety Code for preventing electric shock from induced currents, an assessment of the impact of the proposed action on the potential shock hazard from the transmission lines must be provided.

- 10 CFR 51.70(b). The draft environmental impact statement will be concise, clear, and analytic, and written in plain language with appropriate graphics.... The format provided in Section 1(a) of Appendix A of this subpart should be used. The NRC staff will independently evaluate and be responsible for the reliability of all information used in the draft environmental impact statement.

- 10 CFR 51.71, concerning the draft environmental impact statement, will include a preliminary analysis that considers and weighs the environmental effects of the proposed action, the environmental impacts of alternatives to the proposed action, and alternatives available for reducing or avoiding adverse environmental effects.

- 10 CFR 51.95(c), concerning renewal of an operating license or combined license for a nuclear power plant. Under Parts 52 or 54 of this chapter, the Commission shall prepare an environmental impact statement, which is a supplement to the Commission's NUREG-1437, "Generic Environmental Impact Statement for License Renewal of Nuclear Plants."

- 10 CFR Part 51, Appendix A to Subpart A, para. 7, concerning the environmental consequences of alternatives, including the proposed actions and any mitigating actions which may be taken. Alternatives eliminated from detailed study will be identified and a discussion of those alternatives will be confined to a brief statement of the reasons the alternatives were eliminated. The level of information for each alternative considered in detail will reflect the depth of analysis required for sound decisionmaking.

- 10 CFR Part 51, Appendix B to Subpart A, "Environmental Effect of Renewing the Operating License of a Nuclear Power Plant," Table B-1, "Summary of Findings on NEPA Issues for License Renewal of Nuclear Power Plants"

Technical Rationale

The review conducted under this ESRP leads to the preparation of SEIS sections that incorporate the conclusions in the GEIS related to human health impacts from continued plant operations during the license renewal term, refurbishment, and alternatives to the proposed action. The review should also address any new and significant information.

III. REVIEW PROCEDURES

Suggested steps for the review process are as follows:

1. Review the discussion of potential human health impacts from continued plant operations during the operating license renewal term in the GEIS. This step establishes the basis for evaluating any new and significant human health information identified by the applicant, the public, and the staff. Category 1 and uncategorized human health impact issues identified in the GEIS are listed in the following table.

Table B-1. Summary of Findings on NEPA Issues for License Renewal of Nuclear Power Plants		
Issue	Category	Finding
Human Health		
Radiation exposures to the public	1	SMALL. Radiation doses to the public from continued operations and refurbishment associated with license renewal are expected to continue at current levels, and would be well below regulatory limits.
Radiation exposures to plant workers	1	SMALL. Occupational doses from continued operations and refurbishment associated with license renewal are expected to be within the range of doses experienced during the current license term, and would continue to be well below regulatory limits.
Human health impact from chemicals	1	SMALL. Chemical hazards to plant workers resulting from continued operations and refurbishment associated with license renewal are expected to be minimized by the licensee implementing good industrial hygiene practices as required by permits and Federal and State regulations. Chemical releases to the environment and the potential for impacts to the public are expected to be minimized by adherence to discharge limitations of NPDES and other permits.
Microbiological hazards to plant workers	1	SMALL. Occupational health impacts are expected to be controlled by continued application of accepted industrial hygiene practices to minimize worker exposures as required by permits and Federal and State regulations.

Table B-1. Summary of Findings on NEPA Issues for License Renewal of Nuclear Power Plants		
Issue	**Category**	**Finding**
Chronic effects of electromagnetic fields (EMFs)	N/A	**Uncertain impact.** Studies of 60-Hz EMFs have not uncovered consistent evidence linking harmful effects with field exposures. EMFs are unlike other agents that have a toxic effect (e.g., toxic chemicals and ionizing radiation) in that dramatic acute effects cannot be forced and longer-term effects, if real, are subtle. Because the state of the science is currently inadequate, no generic conclusion on human health impacts is possible.
Physical occupational hazards	1	SMALL. Occupational safety and health hazards are generic to all types of electrical generating stations, including nuclear power plants, and are of small significance if the workers adhere to safety standards and use protective equipment as required by Federal and State Regulations.

The Category 2 human health impact issues identified in the GEIS are listed in the following table.

Table B-1. Summary of Findings on NEPA Issues for License Renewal of Nuclear Power Plants		
Issue	**Category**	**Finding**
Human Health		
Microbiological hazards to the public (plants with cooling ponds or canals or cooling towers that discharge to a river)	2	SMALL, MODERATE, or LARGE. These organisms are not expected to be a problem at most operating plants except possibly at plants using cooling ponds, lakes, or canals, or that discharge into rivers. Impacts would depend on site-specific characteristics.
Electric shock hazards	2	SMALL, MODERATE, or LARGE. Electrical shock potential is of small significance for transmission lines that are operated in adherence with the National Electrical Safety Code (NESC). Without a review of conformance with NESC criteria of each nuclear plant's in-scope transmission lines, it is not possible to determine the significance of the electrical shock potential.

2. Determine whether there is any new human health impact information that should be evaluated. The following sources of information should be included in the search for new information:

- The applicant's ER. An applicant is required by 10 CFR 51.53(c)(3)(iv) to disclose new and significant information on the human health impacts of operating license renewal of which it is aware. In reviewing the applicant's ER, consider the applicant's process for discovering new information and evaluating the significance of any new information. Is the process adequate to ensure a reasonable likelihood that the applicant would be aware of new information?

- Records of public scoping meetings and correspondence related to the operating license renewal application. Compare the human health information presented by the public with information considered in the GEIS. Does the information post date the analysis leading to the GEIS?

- Part 20 standards and regulations. Have the applicable standards and regulations changed since the analysis leading to the GEIS? If so, do these changes affect the NRC evaluation of applications for license renewal?

3. Evaluate the significance of new human health impact information.

4. Review the applicant's ER, including:

   - the applicant's process for identifying new and potentially significant information

   - any new information included in the ER on human health impact issues known to the applicant and the public

5. Prepare a statement for the SEIS that:

   - describes analysis of continued plant operations and refurbishment and summarizes the information that has been reviewed, and the analyses that have been conducted

   - describes measures to mitigate adverse impacts

   - provides the significance level of the environmental impacts

   - describes new and significant information, if any

IV. EVALUATION FINDINGS

The depth and extent of the input to the SEIS would be governed by the extent of the analysis required to reach a conclusion related to the potential human health impacts from continued plant operations and refurbishment. The information that should be included in the SEIS is described in the review procedures.

V. IMPLEMENTATION

The method described in this ESRP would be used by the staff in evaluating conformance with the Commission's regulations, except in those cases in which the applicant for license renewal proposes an acceptable alternative for complying with specified portions of the regulations.

VI. BIBLIOGRAPHY

10 CFR Part 51. *Code of Federal Regulations*, Title 10, *Energy*, Part 51, "Environmental Protection Regulations for Domestic Licensing and Related Regulatory Functions."

Centers for Disease Control and Prevention (CDC). 1996. *Surveillance for Waterborne-Disease Outbreaks—United States, 1993-1994*. M.H. Kramer, B.L. Herwaldt, G.F. Craun, R.L. Calderon, D.D. Juranek. Source: MMWR 45(SS-1): 1–33. April 12.

Institute of Electrical and Electronics Engineers, Inc. (IEEE). 2007. *The National Electrical Safety Code (NESC)*. C2-2007. New York.

National Environmental Policy Act of 1969 (NEPA). 42 USC 4321 et seq.

U.S. Nuclear Regulatory Commission (NRC). 2013. *Generic Environmental Impact Statement for License Renewal of Nuclear Plants*. NUREG-1437, Vols. 1, 2, and 3, Revision 1. Office of Nuclear Reactor Regulation, Washington, D.C.

NUREG-1555, Supplement 1

U.S. NUCLEAR REGULATORY COMMISSION
ENVIRONMENTAL STANDARD
REVIEW PLAN
OFFICE OF NUCLEAR REACTOR REGULATION

4.8.1 MICROBIOLOGICAL HAZARDS TO THE PUBLIC

I. AREAS OF REVIEW

This environmental standard review plan (ESRP) provides guidance for the analysis and assessment of the human-health impacts associated with microbiological hazards to the public associated with heated-water discharges from the plant's cooling system during the renewal term. This issue is identified as a Category 2 issue in the *Generic Environmental Impact Statement for License Renewal of Nuclear Plants* (GEIS; NUREG-1437, Volumes 1, 2, and 3, Revision 1).

The scope includes: (1) review of the impacts to human health from thermophilic microorganisms discussion in the GEIS, (2) evaluation of new information for significance, and (3) preparation of input to the supplemental environmental impact statement (SEIS).

Microorganisms that are associated with cooling towers and thermal discharges can have negative impacts on human health. The presence and numbers of these organisms can be increased by the addition of heat; thus they are called thermophilic organisms. These microorganisms include the enteric pathogens *Salmonella* sp. and *Shigella* sp. as well as *Pseudomonas aeriginosa* and thermophilic fungi. They also include the bacteria *Legionella* sp., which causes Legionnaires' disease, and free-living amoebae of the genera *Naegleria* and *Acanthamoeba*. Exposure to these microorganisms, or in some cases the endotoxins or exotoxins produced by the organisms, can cause illness or death.

Maximum contaminant levels of various microorganisms, including *Legionella*, in public drinking water systems are regulated by 40 CFR 141.70. However, there are no regulations that could be tied to microorganisms that are associated with cooling towers or thermal discharges. Other than the need to

4.8.1-1

USNRC ENVIRONMENTAL STANDARD REVIEW PLAN

Environmental standard review plans are prepared for the guidance of the Office of Nuclear Reactor Regulation staff responsible for environmental reviews for nuclear power plants. These documents are made available to the public as part of the Commission's policy to inform the nuclear industry and the general public of regulatory procedures and policies. Environmental standard review plans are not substitutes for regulatory guides or the Commission's regulations and compliance with them is not required. These supplemental environmental standard review plans are keyed to Regulatory Guide 4.2, Supplement 1, "Preparation of Environmental Reports for Nuclear Power Plant License Renewal Applications."

Published environmental standard review plans will be revised periodically, as appropriate, to accommodate comments and to reflect new information and experience.

Comments and suggestions for improvement will be considered and should be sent to the U.S. Nuclear Regulatory Commission, Office of Nuclear Reactor Regulation, Washington, DC 20555-0001.

assess the impact of thermophilic microorganisms on license renewal, there are no acceptance criteria associated with microbial organisms that may exist in the cooling system and that could affect human health.

Data and Information Needs

The types of data and information needed would be affected by nuclear power plant site- and plant-specific factors. The following data or information may be needed:

- the applicant's environmental report (ER)

- the GEIS

- new information on impacts to human health from thermophilic microorganisms identified by the public and other information sources

II. ACCEPTANCE CRITERIA

Acceptance criteria for the evaluation of human health impacts from thermophilic microorganisms are addressed in ESRP Section 4.8, Human Health.

Technical Rationale

The review conducted under this ESRP leads to the preparation of SEIS sections that incorporate the conclusions in the GEIS related to human health impacts from thermophilic microorganisms associated with continued plant operations during the license renewal term, refurbishment, and alternatives to the proposed action. The review should also address any new and significant information.

III. REVIEW PROCEDURES

Suggested steps for the review process are as follows:

1. Review the discussion of potential impacts to human health from thermophilic microorganisms associated with continued plant operations during the operating license renewal term in the GEIS. This step establishes the basis for evaluating any new and significant socioeconomic information identified by the applicant, the public, and the staff.

2. Review the plant cooling system. If the plant cooling system uses a cooling pond, lake, or canal, or discharges to a river, then continue the analysis at Step 3. Otherwise, prepare a statement for the SEIS that describes the plant cooling system; states that the cooling system discharges to an ocean, a large lake, or other body of water; and concludes that there would not be a detrimental impact from the thermal discharges on the concentration levels of deleterious thermophilic microorganisms.

- a description of the location of the thermal discharges for the plant's cooling system (i.e., a cooling pond, lake, canal) and a characterization of the water body receiving discharges from the cooling system, e.g., a large lake or ocean (from the ER)

- the temperature increase expected for the aquatic environment that is subject to the plant's thermal discharges (from the ER)

- the results of any analyses that have been made for the presence of deleterious thermophilic microorganisms. These include the enteric pathogens *Salmonella* sp. and *Shigella* sp., as well as *Pseudomonas aeriginosa* and thermophilic fungi. In addition, analyses for the presence of unusually high concentrations of the normally present *Legionella* sp. (Legionnaires' disease bacteria) and the free-living amoebae of the genera *Naegleria* and *Acanthamoeba* should be cited (from the ER)

- a list of the outbreaks of waterborne diseases in the United States during the previous 10 years in the vicinity of the plant. This list is published regularly by the Centers for Disease Control and Prevention (CDC 1996)

- an evaluation of available data concerning the occurrence and concentrations of any of the deleterious thermophilic microorganisms listed above in the vicinity of the plant and a determination of whether any of them are present under conditions and in locations that might be harmful to members of the public. If such an evaluation exists, it may be obtained from the applicant or from the State Public Health Department in the State in which the plant is located

3. Consult with the State Public Health Department and review any records associated with waterborne disease outbreaks in the region. If the State Public Health Department is concerned about such outbreaks or the potential for such outbreaks, then continue the analysis at Step 4. Otherwise, prepare a statement for the SEIS describing the plant cooling system that:

- outlines the process leading to the determination that there have been no or few waterborne disease outbreaks in the region

- provides a statement from the State Public Health Department indicating their basis for not being concerned about the potential for an impact to the public health from microbiological organisms associated with the cooling system

- concludes that it appears unlikely that thermal discharges from the plant would increase the number of deleterious thermophilic microorganisms to levels that could cause a public health problem

4. If the State advises that tests should be conducted for concentration of *N. fowleri* (or other deleterious thermophilic microorganisms) in the receiving waters, the licensee should consider performing the tests when the facility has been operating at a power level typical of the level anticipated during the license renewal term for at least a month to ensure a steady state population during the sampling. Samples should be taken at locations of potential public use. An evaluation of the data should be performed and a determination made of the magnitude of potential impacts of *N. fowleri* (or other deleterious thermophilic microorganisms) on public health during the license

renewal term. If the potential for an impact is determined, then continue the analysis at Step 5. If the State does not advise that tests be conducted, but they still have a concern related to the presence of deleterious thermophilic microorganisms, then continue the analysis at Step 5 without the testing. Otherwise, prepare a statement for the SEIS that

- describes the results of the tests that were performed

- provides a statement from the State Public Health Department indicating their basis for not being concerned about the potential for an impact to the public health from microbiological organisms associated with the cooling system as a result of the tests that were performed

- concludes that it appears unlikely that thermal discharges from the plant would increase the number of deleterious thermophilic microorganisms to levels that could cause a public health problem

5. Request that the applicant consider mitigative measures to minimize the potential impacts if the results of the consultation with the State Public Health Department and/or the review of records associated with waterborne disease outbreaks in the region show any cause for concern regarding public health concerns related to deleterious thermophilic microorganisms. Mitigative measures may include:

- setting up and executing a monitoring program for deleterious thermophilic microorganisms

- limiting public access to areas affected by the plant's thermal discharges (such as prohibiting public swimming in the mixing zone of the river)

6. Prepare a statement for the SEIS that

- describes the plant cooling system

- summarizes the information related to any waterborne disease outbreaks in the region

- provides a statement from the State Public Health Department indicating any concerns regarding the potential for an impact to the public health from microbiological organisms associated with the cooling system

- identifies and describes the mitigative measures considered and committed to by the applicant

- concludes that the impacts of microbiological organisms associated with the cooling system are SMALL, MODERATE, or LARGE within the context of the analysis in the GEIS, considering the mitigative measures committed to by the applicant

- describes new and significant information, if any

IV. EVALUATION FINDINGS

The depth and extent of the input to the SEIS would be governed by the extent of the analysis required to reach a conclusion related to the potential impacts on human health from microbiological organisms associated with the plant's cooling system. The information that should be included in the SEIS is described in the review procedures.

V. IMPLEMENTATION

The method described in this ESRP would be used by the staff in evaluating conformance with the Commission's regulations, except in those cases in which the applicant for license renewal proposes an acceptable alternative for complying with specified portions of the regulations.

VI. BIBLIOGRAPHY

The bibliography for this section is provided in ESRP Section 4.8, Human Health.

4.8.2 ELECTRIC SHOCK HAZARDS

I. AREAS OF REVIEW

This environmental standard review plan (ESRP) provides guidance for the review of the electric shock hazards from transmission-line-induced currents. This issue is identified as a Category 2 issue in the *Generic Environmental Impact Statement for License Renewal of Nuclear Plants* (GEIS; NUREG-1437, Volumes 1, 2, and 3, Revision 1).

The scope includes: (1) review of the impacts to human health from electric shock from in-scope transmission-line-induced currents in the GEIS, (2) evaluation of new information for significance, and (3) preparation of input to the supplemental environmental impact statement (SEIS).

The scope should include determining if transmission lines constructed for the purpose of connecting the plant to the transmission system meet the recommendations of the National Electrical Safety Code (NESC) for preventing electric shock from induced currents. If not, the scope includes assessing the impact of the proposed action on the potential shock hazard from the transmission lines. The scope also includes preparation of input to the SEIS.

Data and Information Needs

The types of data and information needed would be affected by nuclear power plant site- and plant-specific factors. The following data or information may be needed:

4.8.2-1

USNRC ENVIRONMENTAL STANDARD REVIEW PLAN

Environmental standard review plans are prepared for the guidance of the Office of Nuclear Reactor Regulation staff responsible for environmental reviews for nuclear power plants. These documents are made available to the public as part of the Commission's policy to inform the nuclear industry and the general public of regulatory procedures and policies. Environmental standard review plans are not substitutes for regulatory guides or the Commission's regulations and compliance with them is not required. These supplemental environmental standard review plans are keyed to Regulatory Guide 4.2, Supplement 1, "Preparation of Environmental Reports for Nuclear Power Plant License Renewal Applications."

Published environmental standard review plans will be revised periodically, as appropriate, to accommodate comments and to reflect new information and experience.

Comments and suggestions for improvement will be considered and should be sent to the U.S. Nuclear Regulatory Commission, Office of Nuclear Reactor Regulation, Washington, DC 20555-0001.

- the applicant's environmental report (ER)

- the GEIS

- new information on impacts to human health from electric shock from transmission-line-induced currents identified by the public and other information sources

II. ACCEPTANCE CRITERIA

Acceptance criteria for the evaluation of electric shock from transmission-line-induced currents are addressed in ESRP Section 4.8, Human Health, with the following addition:

- 10 CFR 51.53(c)(3)(ii)(H), concerning assessing impacts of transmission systems not meeting NESC criteria

Additional regulatory positions and specific criteria in support of the regulation identified above are as follows:

- NESC (IEEE 2007) provides guidance concerning shock hazards

Technical Rationale

The review conducted under this ESRP leads to the preparation of SEIS sections that incorporate the conclusions in the GEIS related to human health impacts from electric shock from transmission-line-induced currents associated with continued plant operations during the license renewal term, refurbishment, and alternatives to the proposed action. The review should also address any new and significant information.

III. REVIEW PROCEDURES

Suggested steps for the review process are as follows:

1. Review the discussion of the issues associated with electric shock hazards from induced transmission line currents in the GEIS.

2. Review the route of the in-scope transmission lines.

3. Review the applicant's analysis demonstrating that the transmission lines continue to meet NESC clearance standards to which they were built.
   - The following data or information may be needed to assess human health impacts from electric shock from transmission-line-induced currents:
     - a description of the in-scope transmission lines

- verification of initial transmission line conformance with NESC criteria (NESC edition to which the lines were built or a later edition)

- a description of a transmission line management program, if any, including continued compliance with NESC electrical shock provisions

- plans to bring lines into conformance with NESC criteria if not already in compliance. Consider basic electrical design parameters, including transmission design voltage or voltages, line capacity, conductor type and configuration, spacing between phases, minimum conductor clearances to ground, maximum predicted electric field strength(s) at 1 meter above ground, the predicted electric field strength(s) at the edge of the right-of-way in kilovolts per meter (kV/m), and the design bases for these values (from the ER)

- (if NESC clearance standards cannot be demonstrated) a transmission line survey identifying sites or areas that do not meet the standards and that may not meet the standards following anticipated changes in transmission-line operations or changes in land use in the right-of-way

- If the applicant does not state that in-scope transmission lines meet electrical shock hazard of the NESC code or the applicant's demonstration is not adequate, then continue the review at Step 4. Otherwise, prepare a statement for the SEIS that

    - describes the route of the in-scope transmission lines

    - describes the line (voltage, capacity, conductor configuration, minimum conductor-to-ground clearance, and maximum predicted electrical field strengths 1 meter above ground, etc.)

    - provides the basis for the staff evaluation

    - concludes that the system meets the criteria of the NESC

4. Identify any sites or areas where the transmission lines fail to meet the NESC clearance standards. These areas should be shown on maps, photographs, or drawings to be included in the SEIS.

5. Identify measures that could be taken to meet the standards in the areas where the transmission lines fail to meet the NESC standards. Determine which measures the applicant plans or proposes to undertake, if any, and whether those measures would result in transmission lines meeting the standards.

6. Identify and evaluate mitigation measures for those areas where the transmission lines would not meet NESC standards.

7. Prepare a statement for the SEIS that:

- describes the route of the in-scope transmission lines

- describes the line (voltage, capacity, conductor configuration, minimum conductor-to-ground clearance, and maximum predicted electrical field strengths 1 meter above ground, etc.) and potential shock hazard from the transmission lines

- identifies sites or areas where NESC standards would not be met and explains why the standards are not appropriate to the situation or why the applicant would not make modifications to meet standards

- describes measures to mitigate potential impacts in those areas

- provides the significance level of the environmental impacts

- describes new and significant information, if any

IV. EVALUATION FINDINGS

The depth and extent of the input to the SEIS would be governed by the extent of the analysis required to reach a conclusion related to the potential electric shock from transmission-line-induced currents. The information that should be included in the SEIS is described in the review procedures.

V. IMPLEMENTATION

The method described in this ESRP would be used by the staff in evaluating conformance with the Commission's regulations, except in those cases in which the applicant for license renewal proposes an acceptable alternative for complying with specified portions of the regulations.

VI. BIBLIOGRAPHY

The bibliography for this section is provided in ESRP Section 4.8, Human Health.

U.S. NUCLEAR REGULATORY COMMISSION

ENVIRONMENTAL STANDARD
REVIEW PLAN

OFFICE OF NUCLEAR REACTOR REGULATION

4.9 ENVIRONMENTAL JUSTICE

On February 11, 1994, the President signed Executive Order 12898 "Federal Actions to Address Environmental Justice in Minority Populations and Low-Income Populations," which directs all Federal agencies to develop strategies for considering environmental justice in their programs, policies, and activities. Environmental justice is described in the Executive Order as "identifying and addressing, as appropriate, disproportionately high and adverse human health or environmental effects of its programs, policies, and activities on minority populations and low-income populations." On December 10, 1997, the Council on Environmental Quality (CEQ) issued "Environmental Justice Guidance Under the National Environmental Policy Act." The Council developed this guidance to, "further assist Federal agencies with their National Environmental Policy Act (NEPA) procedures."

On August 24, 2004, the Commission issued a "Policy Statement on the Treatment of Environmental Justice Matters in NRC Regulatory and Licensing Actions" (69 FR 52040), which states, "the Commission is committed to the general goals set forth in E.O. 12898, and strives to meet those goals as part of its NEPA review process." The following guidance is consistent with this policy statement.

I. AREAS OF REVIEW

This environmental standard review plan (ESRP) provides guidance on conducting environmental justice reviews for proposed actions requiring an environmental impact statement (EIS) as part of NRC's compliance with NEPA. This issue is identified as a Category 2 issue in the *Generic Environmental Impact Statement for License Renewal of Nuclear Plants* (GEIS; NUREG-1437, Volumes 1, 2, and 3, Revision 1). The scope includes: (1) review of the impacts to minority and low-income populations in the applicant's environmental report (ER) and GEIS, (2) evaluation of new information for significance, and (3) preparation of input to the supplemental environmental impact statement (SEIS).

Guidelines for specific information requirements for environmental justice determinations are described in NRR Office Instruction LIC-203, Revision 2: "Procedural Guidance for Preparing Environmental Assessments and Considering Environmental Issues" (NRC 2009). This office instruction is revised periodically. For current guidance, obtain the latest revision.

The scope should include an analysis of the potential effects from continued power plant operations during the renewal term and refurbishment associated with license renewal on minority and low-income populations. This review should be of sufficient detail to permit the determination of whether these effects are likely to be disproportionately high and adverse to minority and low-income populations.

The staff should consider the demographic composition of the affected area to determine the location of minority and low-income populations and whether they may be affected by the proposed action. The staff then needs to determine if human health or environmental effects would have a disproportionately high and adverse effect on minority or low-income populations.

Data and Information Needs

The types of data and information needed would be affected by nuclear power plant site- and plant-specific factors. The following data or information may be needed:

- the applicant's ER

- the GEIS

- new information on environmental justice identified by the public and other information sources

II. ACCEPTANCE CRITERIA

The acceptance criteria for environmental justice impacts are based on the relevant requirements of the following:

- Executive Order 12898 (59 FR 7629) provides guidance concerning Federal actions to address environmental justice in minority and low-income populations.
- "Policy Statement on the Treatment of Environmental Justice Matters in NRC Regulatory and Licensing Actions," (69 FR 52040) affirms that the NRC is committed to the general goals of Executive Order 12898 and states that the NRC strives to meet those goals as part of its NEPA review process for licensing actions.

- 10 CFR 51.45(c), "Analysis." The environmental report must include an analysis that considers and balances the environmental effects of the proposed action, the environmental impacts of alternatives to the proposed action, and alternatives available for reducing or avoiding adverse environmental effects.

- 10 CFR 51.53(c)(2). The report must contain a description of the proposed action, including the applicant's plans to modify the facility or its administrative control procedures as described in accordance with 10 CFR 54.21 of this chapter. This report must describe in detail the affected environment around the plant, the modifications directly affecting the environment or any plant effluents, and any planned refurbishment activities. In addition, the applicant shall discuss in this report the environmental impacts of alternatives and any other matters discussed in 10 CFR 51.45.

- 10 CFR 51.53(c)(3)(ii)(N). Applicants shall provide information on the general demographic composition of minority and low-income populations and communities (by race and ethnicity) residing in the immediate vicinity of the plant that could be affected by the renewal of the plant's operating license, including any planned refurbishment activities, and ongoing and future plant operations.

- 10 CFR 51.70(b). The draft environmental impact statement will be concise, clear, and analytic, and written in plain language with appropriate graphics….The format provided in Section 1(a) of Appendix A of this subpart should be used. The NRC staff will independently evaluate and be responsible for the reliability of all information used in the draft environmental impact statement.

- 10 CFR 51.71(d), concerning the draft environmental impact statement, will include a preliminary analysis that considers and weighs the environmental effects of the proposed action; the environmental impacts of alternatives to the proposed action; and alternatives available for reducing or avoiding adverse environmental effects.

- 10 CFR 51.95(c), concerning renewal of an operating license or combined license for a nuclear power plant. Under Parts 52 or 54 of this chapter, the Commission shall prepare an environmental impact statement, which is a supplement to the Commission's NUREG-1437, "Generic Environmental Impact Statement for License Renewal of Nuclear Plants."

- 10 CFR 51, Appendix A to Subpart A (7), concerning the environmental consequences of alternatives, including the proposed actions and any mitigating actions which may be taken. Alternatives eliminated from detailed study will be identified and a discussion of those alternatives will be confined to a brief statement of the reasons why the alternatives were eliminated. The level of information for each alternative considered in detail will reflect the depth of analysis required for sound decisionmaking.

- 10 CFR Part 51, Appendix B to Subpart A, "Environmental Effect of Renewing the Operating License of a Nuclear Power Plant," Table B-1, "Summary of Findings on NEPA Issues for License Renewal of Nuclear Power Plants"

Additional regulatory positions and specific criteria in support of the regulations identified above are as follows:

- Council on Environmental Quality (CEQ) guidance for addressing environmental justice, "Environmental Justice: Guidance under the National Environmental Policy Act," December 10, 1997 (CEQ 1997)

- Guidelines for specific information requirements for environmental justice determinations are described in NRR Office Instruction LIC-203, Revision 2, "Procedural Guidance for Preparing Environmental Assessments and Considering Environmental Issues." NRR Office Instruction LIC-203 is revised periodically. Obtain a copy of the latest revision for current guidance.

Technical Rationale

The review conducted under this ESRP leads to the preparation of SEIS sections that incorporate the conclusions in the GEIS related to environmental justice impacts from continued plant operations during the license renewal term, refurbishment associated with license renewal, and alternatives to the proposed action. The review should also address any new and significant information.

III. REVIEW PROCEDURES

The review procedure should be as follows:

1. Review the discussion of potential environmental justice impacts from continued plant operations during the license renewal term in the GEIS. This step establishes the basis for evaluating any new and significant socioeconomic information identified by the applicant, the public, and NRC staff. Category 2 environmental justice impact issue and finding is provided in the following table.

Table B-1. Summary of Findings on NEPA Issues for License Renewal of Nuclear Power Plants		
Issue	Category	Finding
Environmental Justice		
Minority and low-income populations	2	Impacts to minority and low-income populations and subsistence consumption resulting from continued operations and refurbishment associated with license renewal will be addressed in plant-specific reviews. See NRC Policy Statement on the Treatment of Environmental Justice Matters in NRC Regulatory and Licensing Actions (69 FR 52040; August 24, 2004).

2. Identify minority and low-income populations within a 50-mile (80-kilometer) radius of the nuclear plant. For each census block group within this area, minority and low-income populations are identified when (1) the minority or low-income population of an impacted area exceeds 50 percent or (2) the minority or low-income population percentage of the impacted area is meaningfully greater than the minority or low-income population percentage in the general population or other appropriate unit of geographic analysis (e.g., 50-mile radius geographic area or county). All block groups with minority and low-income percentages higher than the geographic area should be identified on 50-mile radius maps.

3. Identify environmental justice issues and unique characteristics of minority and low-income populations/communities during the scoping process:

- Determine geographic distribution by race, ethnicity, and poverty, as well as delineation of Tribal lands. Identify any unique characteristics of minority and low-income populations and the "special character" of communities located near the nuclear plant.

- In calculating the minority populations, individual(s) who are members of the following population groups are considered minority individuals:

  - Race (Not Hispanic or Latino):
    Black or African American
    American Indian or Alaska Native
    Asian
    Native Hawaiian and Other Pacific Islander
    Some other race
    Two or more races
  - Ethnicity:
    Hispanic or Latino (of any race)

- Low-income population is defined as individuals or families living below the poverty level as defined by the U.S. Census Bureau (e.g., the U.S. Census Bureau's Current Population Reports, Series P-60 on Income and Poverty).

- Sources of information for determining geographic distribution and location of minority populations:

  - EJView online geographic information system (GIS) tool offered by the U.S. Environmental Protection Agency

  - Local governments

  - State agencies

  - Local universities

4. Determine whether license renewal would have any human health and environmental effects on minority or low-income populations or whether there are other environmental justice concerns.

- Potential human health and environmental effects are determined through NRC's NEPA review process:

  - Impacts that could potentially affect or cause concern to minority and low-income populations are evaluated for other environmental resource areas (e.g., air and water quality, socioeconomics, and cultural resources) during the license renewal environmental review. Any potential effects and/or concerns should be summarized in the environmental justice section of the SEIS.

  - In considering human health and environmental impacts to minority and low-income populations, different patterns of consumption of natural resources should also be considered (i.e., differences in rates and/or pattern of fish, vegetable, water, and/or wildlife consumption reflective of the unique characteristics of minority and low-income populations and the "special character" of communities located near the nuclear plant) (see Section 4-4 of Executive Order 12898, "Subsistence Consumption of Fish and Wildlife").

Consider whether there are any means or pathways for minority or low-income populations to be disproportionately affected by license renewal-related activities. Examine the potential impacts to special pathway receptors (e.g., American Indian, Hispanic, and others living a traditional lifestyle pattern of subsistence). For example, special pathway impacts take into account levels of contaminants in native vegetation, crops, soils and sediments, surface water, fish, and game animals in the vicinity of nuclear plant sites.

- Sources of information include:

  - Radiological Environmental Monitoring Program (REMP), annual radiological environmental operating reports
  - State radiological monitoring programs

5. Determine whether there are disproportionately high and adverse human health or environmental effects on minority and low-income populations.

   - Consider the following questions:

     - Would the impact(s) be greater for minority and low-income populations than the general population?

     - Are there any unique effects experienced by minority and low-income populations that would not be experienced by the general population?

   - After identifying human health and environmental impacts that could disproportionately affect minority and low-income populations, it is necessary to determine if the effect(s) would be high and adverse. Another way of stating this:

     - Would the effect(s) on minority and low-income populations be significant, unacceptable, or above generally accepted norms such as regulatory limits or State and local statutes and ordinances? Each human health and environmental impact, and where appropriate, the cumulative and multiple effects of the impact(s), should be reviewed for significance.

   - To the extent practicable, mitigation measures should reflect the needs and preferences of the affected minority or low-income populations and communities.

6. Make a determination regarding impacts to minority and low-income populations and document the conclusion in the SEIS.

IV. EVALUATION FINDINGS

The extent of information and the depth of analysis conducted for the SEIS would be governed by the significance and magnitude of the potential effects from continued operation of the nuclear power plant during the renewal term and related refurbishment activities on minority and low-income populations.

V. IMPLEMENTATION

The review plan guidance described in this ESRP would be used by the staff in conformance with the Commission's environmental protection regulations in 10 CFR Part 51 and the "Policy Statement on the Treatment of Environmental Justice Matters in NRC Regulatory and Licensing Actions."

VI. BIBLIOGRAPHY

10 CFR Part 51. *Code of Federal Regulations*, Title 10, *Energy,* Part 51, "Environmental Protection Regulations for Domestic Licensing and Related Regulatory Functions."

69 FR 52040. U.S. Nuclear Regulatory Commission. Policy Statement on the Treatment of Environmental Justice Matters in NRC Regulatory and Licensing Actions. 2004.

Council on Environmental Quality (CEQ). 1997. *Environmental Justice: Guidance under the National Environmental Policy Act*. Washington, D.C. December 10. Available URL: http://www.nepa.gov/nepa/regs/ej/justice.pdf.

Executive Order 12898. 1994. *Federal Actions to Address Environmental Justice in Minority and Low-Income Populations*. 59 FR 7629.

National Environmental Policy Act of 1969 (NEPA). 42 USC 4321 et seq.

U.S. Nuclear Regulatory Commission (NRC). 2009. *Procedural Guidance for Preparing Environmental Assessments and Considering Environmental Issues*. LIC-203, Revision 2. Office of Nuclear Reactor Regulation. ADAMS Accession Number ML080840323.

U.S. Nuclear Regulatory Commission (NRC). 2013. *Generic Environmental Impact Statement for License Renewal of Nuclear Plants*. NUREG-1437, Vols. 1, 2, and 3, Revision 1. Office of Nuclear Reactor Regulation, Washington, D.C.

U.S. NUCLEAR REGULATORY COMMISSION

ENVIRONMENTAL STANDARD REVIEW PLAN

OFFICE OF NUCLEAR REACTOR REGULATION

4.10 WASTE MANAGEMENT

I. AREAS OF REVIEW

This environmental standard review plan (ESRP) provides guidance for the review of waste management activities at nuclear power plants during the license renewal term and refurbishment.

The scope includes (1) review of the discussion of waste management during the license renewal term in the GEIS, (2) identification and evaluation of any new information, and (3) preparation of input to the supplemental environmental impact statement (SEIS).

Data and Information Needs

The types of data and information needed would be affected by nuclear power plant site- and plant-specific factors. The following data or information may be needed:

* a description of the applicant's process for identifying new and potentially significant information

* any new information included in the ER on waste management and pollution prevention at the plant

* the GEIS

USNRC ENVIRONMENTAL STANDARD REVIEW PLAN

Environmental standard review plans are prepared for the guidance of the Office of Nuclear Reactor Regulation staff responsible for environmental reviews for nuclear power plants. These documents are made available to the public as part of the Commission's policy to inform the nuclear industry and the general public of regulatory procedures and policies. Environmental standard review plans are not substitutes for regulatory guides or the Commission's regulations and compliance with them is not required. These supplemental environmental standard review plans are keyed to Regulatory Guide 4.2, Supplement 1, "Preparation of Environmental Reports for Nuclear Power Plant License Renewal Applications."

Published environmental standard review plans will be revised periodically, as appropriate, to accommodate comments and to reflect new information and experience.

Comments and suggestions for improvement will be considered and should be sent to the U.S. Nuclear Regulatory Commission, Office of Nuclear Reactor Regulation, Washington, DC 20555-0001.

II. ACCEPTANCE CRITERIA

Acceptance criteria for the evaluation of waste management during the license renewal term are based on the relevant requirements of the following regulations:

- 10 CFR 51.70(b), concerning independent evaluation and responsibility for the reliability of information used in the draft SEIS

- 10 CFR 51.95(c)(4), concerning NRC staff's obligation to consider significant new information

Technical Rationale

The technical rationale for evaluating the renewal term waste management is discussed in the following paragraph:

The NRC staff is required by 10 CFR 51.95(c)(4) to integrate conclusions, as amplified by the supporting information in the GEIS, for issues designated as Category 1 or resolved Category 2 information developed for those open Category 2 issues applicable to the plant, and any significant new information. The review conducted under this ESRP leads to preparation of sections of the SEIS that incorporate the conclusions in the GEIS related to solid waste management during the license renewal term, as appropriate, and addresses significant new information, if any.

III. REVIEW PROCEDURES

Suggested steps for the review process are as follows:

1. Review the discussion of waste management during the license renewal term in the GEIS to identify the information considered and the conclusions reached. This step establishes the base for evaluation of information identified by the applicant, the public, and the staff. The table on the following page lists the renewal term waste management issues considered in the GEIS.

2. Determine if there is new information on these issues that should be evaluated. The following sources of information should be included in the search for new information:

    - the applicant's ER. An applicant is required by 10 CFR 51.53(c)(3)(iv) to disclose new and significant information regarding the environmental impacts of license renewal of which it is aware. In reviewing the applicant's ER, consider the applicant's process for discovering new information and evaluating the significance of any new information discovered.

    - records of public meetings and correspondence related to the application. Compare information presented by the public with information considered in the GEIS.

    If the search conducted in this step reveals new information, continue with the analysis. Otherwise, prepare the section for the SEIS describing the search for new information, stating the conclusion that there is no new information, and adopting the conclusions from the GEIS.

Table B-1. Summary of Findings on NEPA Issues for License Renewal of Nuclear Power Plants		
Issue	**Category**	**Finding**
Waste Management		
Low-level waste storage and disposal	1	SMALL. The comprehensive regulatory controls that are in place and the low public doses being achieved at reactors ensure that the radiological impacts to the environment would remain small during the license renewal term.
Onsite storage of spent nuclear fuel	1	SMALL. The expected increase in the volume of spent fuel from an additional 20 years of operation can be safely accommodated onsite during the license renewal term with small environmental effects through dry or pool storage at all plants.
Offsite radiological impacts of spent nuclear fuel and high-level waste disposal	N/A	UNCERTAIN IMPACT. The generic conclusion on offsite radiological impacts of spent nuclear fuel and high-level waste is not being finalized pending the completion of a generic environmental impact statement on waste confidence.[*]
Mixed-waste storage and disposal	1	SMALL. The comprehensive regulatory controls and the facilities and procedures that are in place ensure proper handling and storage, as well as negligible doses and exposure to toxic materials for the public and the environment at all plants. License renewal would not increase the small, continuing risk to human health and the environment posed by mixed waste at all plants. The radiological and nonradiological environmental impacts of long-term disposal of mixed waste from any individual plant at licensed sites are small.
Nonradioactive waste storage and disposal	1	SMALL. No changes to systems that generate nonradioactive waste are anticipated during the license renewal term. Facilities and procedures are in place to ensure continued proper handling, storage, and disposal, as well as negligible exposure to toxic materials for the public and the environment at all plants.

* As a result of the decision of United States Court of Appeals in *New York v. NRC*, 681 F.3d 471 (D.C. Cir. 2012), the NRC cannot rely upon its waste confidence decision and rule until it has taken those actions that will address the deficiencies identified by the D.C. Circuit. Although the waste confidence decision and rule did not assess the impacts associated with disposal of spent nuclear fuel and high-level waste in a repository, it did reflect the Commission's confidence, at the time, in the technical feasibility of a repository and when that repository could have been expected to become available. Without the analysis in the waste confidence decision and rule regarding the technical feasibility and availability of a repository, the NRC cannot assess how long the spent fuel will need to be stored onsite.

3. Evaluate the significance of new information.

4. Prepare a section for the SEIS describing the search for new information, summarizing new information found, presenting results of evaluation of significance, and adopting conclusions from the GEIS modified as necessary to account for significant new information.

IV. EVALUATION FINDINGS

The depth and extent of the input to the SEIS would be determined by the analysis required to reach a conclusion related to waste management and pollution prevention during the license renewal term. The information that should be included in the SEIS is described in the review procedures.

V. IMPLEMENTATION

The method described in this ESRP would be used by the staff in evaluating conformance with the Commission's regulations, except in those cases in which the applicant for license renewal proposes an acceptable alternative for complying with specified portions of the regulations.

VI. BIBLIOGRAPHY

10 CFR Part 51. *Code of Federal Regulations*, Title 10, *Energy,* Part 51, "Environmental Protection Regulations for Domestic Licensing and Related Regulatory Functions."

National Environmental Policy Act of 1969 (NEPA). 42 USC 4321 et seq.

U.S. Nuclear Regulatory Commission (NRC). 2013. *Generic Environmental Impact Statement for License Renewal of Nuclear Plants.* NUREG-1437, Vols. 1, 2, and 3, Revision 1. Office of Nuclear Reactor Regulation, Washington, D.C.

4.11 IMPACTS COMMON TO ALL ALTERNATIVES

4.11.1 ENVIRONMENTAL CONSEQUENCES OF THE URANIUM FUEL CYCLE

I. AREAS OF REVIEW

This environmental standard review plan (ESRP) provides guidance for the preparation of introductory paragraphs for the portion of the supplemental environmental impact statement (SEIS) that describes environmental impacts of the uranium fuel cycle during the renewal term.

The scope includes (1) review of the discussion of the uranium fuel cycle in the GEIS, (2) identification and evaluation of new information related to the uranium fuel cycle, (3) preparation of input to the SEIS that presents the analyses related to those Category 1 issues.

Data and Information Needs

The types of data and information needed would be affected by nuclear power plant site- and plant-specific factors. The following data or information may be needed:

- a description of the applicant's process for identifying new and potentially significant information on environmental issues related to the uranium fuel cycle during the renewal term

- new information on the uranium fuel cycle during the renewal term known to the applicant

- new and potentially significant information on the uranium fuel cycle identified by the public

4.11.1-1

- a list of environmental issues related to the uranium fuel cycle during the renewal term for which there is significant new information

II. ACCEPTANCE CRITERIA

Acceptance criteria for the evaluation of the uranium fuel cycle are based on the relevant requirements of the following regulations:

- 10 CFR 51.45(b) concerning environmental considerations in the applicant's ER

- 10 CFR 51.45(d) concerning discussion of compliance with applicable environmental quality standards and requirements in the applicant's ER

- 10 CFR 51.53(c)(3)(ii) concerning analyses required in ERs submitted at the license renewal stage

- 10 CFR 51.70(b) concerning an independent evaluation of the assessment and the reliability of information used in the assessment

- 10 CFR 51.71(d) concerning compliance with environmental quality standards and requirements that have been imposed by Federal, State, regional, local, and affected American Indian Tribal agencies

- 10 CFR 51.95(c)(4) concerning consideration of significant new information

Technical Rationale

The NRC staff is required by 10 CFR 51.95(c)(4) to integrate conclusions, as amplified by the supporting information in the GEIS, for issues designated as Category 1 (with the exception of offsite radiological impacts—collective impacts from other than the disposal of spent fuel and high-level waste) or resolved Category 2, information developed for those open Category 2 issues applicable to the plant, and any significant new information in an EIS prepared at the license renewal stage. The review conducted under this ESRP leads to preparation of sections of the SEIS that incorporate the conclusions from the GEIS related to the uranium fuel cycle and address significant new information, if any.

III. REVIEW PROCEDURES

Suggested steps for the review process are as follows:

1. Review the discussion of the issue in the GEIS to identify the information considered and the conclusions reached. This step establishes the base for evaluation of information identified by the applicant, the public, and the staff. The following table lists the uranium fuel cycle issues that were addressed in the GEIS for which generic conclusions were reached.

Table B-1. Summary of Findings on NEPA Issues for License Renewal of Nuclear Power Plants		
Issue	Category	Finding
Uranium Fuel Cycle		
Offsite radiological impacts – individual impacts from other than the disposal of spent fuel and high-level waste	1	SMALL. The impacts to the public from radiological exposures have been considered by the Commission in Table S-3 of this part. Based on information in the GEIS, impacts to individuals from radioactive gaseous and liquid releases, including radon-222 and technetium-99, would remain at or below the NRC's regulatory limits.
Offsite radiological impacts – collective impacts from other than the disposal of spent fuel and high-level waste	1	There are no regulatory limits applicable to collective doses to the general public from fuel-cycle facilities. The practice of estimating health effects on the basis of collective doses may not be meaningful. All fuel-cycle facilities are designed and operated to meet the applicable regulatory limits and standards. The Commission concludes that the collective impacts are acceptable. The Commission concludes that the impacts would not be sufficiently large to require the NEPA conclusion, for any plant, that the option of extended operation under 10 CFR Part 54 should be eliminated. Accordingly, while the Commission has not assigned a single level of significance for the collective impacts of the uranium fuel cycle, this issue is considered Category 1.
Nonradiological impacts of the uranium fuel cycle	1	SMALL. The nonradiological impacts of the uranium fuel cycle resulting from the renewal of an operating license for any plant would be small.
Transportation	1	SMALL. The impacts of transporting materials to and from uranium-fuel-cycle facilities on workers, the public, and the environment are expected to be small.

2. Determine if there is new information on this issue that should be evaluated. The following sources of information should be included in the search for new information:

  • the applicant's ER. When reviewing, consider the applicant's process for discovering new information and evaluating the significance of any new information discovered

  • records of public meetings and correspondence related to the application

  • environmental quality standards and regulations

If the search conducted in this step reveals new information, then continue with Step 3.

3. Evaluate the significance of new information. If new information is significant, prepare a concise statement(s) of the issues raised by significant new information and provide these statements to the Project Manager (PM) for review and disposition.

IV. EVALUATION FINDINGS

The depth and extent of the input to the SEIS would be governed by the extent of the analysis required to reach conclusions on issues related to the uranium fuel cycle during the renewal term. The information that should be included in the SEIS is described in the review procedures.

V. IMPLEMENTATION

The method described in this ESRP would be used by the staff in evaluating conformance with the Commission's regulations, except in those cases in which the applicant for license renewal proposes an acceptable alternative for complying with specified portions of the regulations.

VI. BIBLIOGRAPHY

10 CFR Part 51. *Code of Federal Regulations*, Title 10, *Energy,* Part 51, "Environmental Protection Regulations for Domestic Licensing and Related Regulatory Functions."

10 CFR Part 54. *Code of Federal Regulations*, Title 10, *Energy,* Part 54, "Requirements for Renewal of Operating Licenses for Nuclear Power Plants."

National Environmental Policy Act of 1969 (NEPA). 42 USC 4321 et seq.

U.S. Nuclear Regulatory Commission (NRC). 2013. *Generic Environmental Impact Statement* (GEIS) *for License Renewal of Nuclear Plants.* NUREG-1437, Vols. 1, 2, and 3, Revision 1. Office of Nuclear Reactor Regulation, Washington, D.C.

U.S. NUCLEAR REGULATORY COMMISSION

ENVIRONMENTAL STANDARD REVIEW PLAN

OFFICE OF NUCLEAR REACTOR REGULATION

4.11.2 ENVIRONMENTAL CONSEQUENCES OF REPLACEMENT POWER ALTERNATIVE FUEL CYCLES

I. AREAS OF REVIEW

This environmental standard review plan (ESRP) provides guidance for the review of the environmental impacts of replacement power alternative fuel cycles during the renewal term.

The scope includes (1) review of the discussion of potential impacts of replacement power alternative fuel cycles in the GEIS, (2) identification and evaluation of new information related to potential impacts of replacement power alternative fuel cycles, (3) preparation of input to the SEIS that presents the analyses related to those Category 1 issues.

Data and Information Needs

The reviewer for this ESRP may obtain the following information from the Environmental Project Manager:

- organizational structure of the SEIS
- list of environmental impacts associated with replacement power alternative fuel cycles that have been determined to be inapplicable to the applicant's plant and the reason for each determination

II. ACCEPTANCE CRITERIA

Acceptance criteria for the review of environmental impacts are based on the relevant requirements of the following regulations:

USNRC ENVIRONMENTAL STANDARD REVIEW PLAN

Environmental standard review plans are prepared for the guidance of the Office of Nuclear Reactor Regulation staff responsible for environmental reviews for nuclear power plants. These documents are made available to the public as part of the Commission's policy to inform the nuclear industry and the general public of regulatory procedures and policies. Environmental standard review plans are not substitutes for regulatory guides or the Commission's regulations and compliance with them is not required. These supplemental environmental standard review plans are keyed to Regulatory Guide 4.2, Supplement 1, "Preparation of Environmental Reports for Nuclear Power Plant License Renewal Applications."

Published environmental standard review plans will be revised periodically, as appropriate, to accommodate comments and to reflect new information and experience.

Comments and suggestions for improvement will be considered and should be sent to the U.S. Nuclear Regulatory Commission, Office of Nuclear Reactor Regulation, Washington, DC 20555-0001.

- 10 CFR 51.53(c), concerning the content of environmental reports (ERs) and required analyses

- 10 CFR 51.70(b), concerning preparation of a draft environmental impact statement (EIS) that is concise, clear, analytic, and written in plain language

- 10 CFR 51.71, concerning preparation of draft EIS at the license renewal stage

- 10 CFR 51.95(c), concerning preparation of a final EIS at the license renewal stage

- 10 CFR 51, Subpart A, Appendix B, concerning the Commission's findings on the scope and magnitude of environmental impacts renewing the operating license for a nuclear power plant

Technical Rationale

The GEIS does not contain any conclusions regarding the environmental impact or acceptability of alternatives to license renewal. Accordingly, the NRC must conduct an analysis of reasonable alternatives to license renewal in plant-specific environmental reviews. A reasonable alternative must be commercially viable on a utility scale and operational prior to the expiration of the reactor's operating license or expected to become commercially viable on a utility scale and operational prior to the expiration of the reactor's operating license. This ESRP examines the potential environmental impacts associated with the replacement power alternative fuel cycles. If a renewed license were denied, then the plant generally would be decommissioned earlier than if the license were renewed, and other electric-generating sources would be pursued if power were still needed.

Analysis of replacement power alternative fuel cycles does not involve the determination of whether any power is needed or should be generated. The decision to generate power and the determination of how much power is needed are at the discretion of State and utility officials.

The potential environmental impacts evaluated include land use, ecology, aesthetics, water quality, air quality, waste management, human health, socioeconomics, and historic and cultural resources.

III. REVIEW PROCEDURES

To analyze the environmental impact of replacement power alternative fuel cycles, the reviewer should complete the following steps:

1. Review the discussion of potential environmental impacts of replacement power alternative fuel cycles in the GEIS to identify the information considered and the conclusions reached. This step establishes the base for evaluation of information identified by the applicant, the public, and the staff.

2. Obtain information for evaluation. The following sources of information should be included in the search for information:

- the applicant's ER. An applicant is required by 10 CFR 51.53(c)(3)(iv) to disclose new and significant information regarding the environmental impacts of license renewal of which it is aware. In reviewing the applicant's ER, consider the applicant's process for discovering new information and evaluating the significance of any new information discovered.

- records of public meetings and correspondence related to the application. Compare information presented by the public with information considered in the GEIS.

3. Determine, from the scope of environmental impacts of replacement power alternative fuel cycles, those that are minor and those that are likely to be sufficiently important to require detailed analysis.

   If, based on this analysis, the reviewer determines that there would be more than minor impacts, proceed to Step 4. Otherwise, if the reviewer determines that there would be no environmental impacts or that the impacts would be minor, develop a statement to this effect.

3. Analyze the environmental impacts associated with replacement power alternative fuel cycles, as follows:

   - Identify and calculate the likely environmental impacts of required replacement power alternative fuel cycles including conservation and purchased or imported power, based on the GEIS, the applicant's ER, and the integrated resource plans for the area(s) or region(s) currently or (if different) likely to be served by the plant. Assume appropriate mitigation measures (for example emission control technologies and best management practices) for each replacement power alternative.

   - Describe the impacts in sufficient detail so that reviewers may compare the adverse and beneficial impacts of the alternatives with those of renewing the operating license. Impact analyses should consider land use, water quality, air quality, ecological resources, human health, social and economic systems, waste management, aesthetics, and cultural resources. The impacts analyses should include direct, indirect, and cumulative impacts. For each alternative, the analysis should identify and, to the extent possible, quantify, unavoidable adverse impacts, irreversible and irretrievable resource commitments, and tradeoffs between short-term use and long-term productivity of the environment. To the extent possible, each alternative should be analyzed on a nuclear power plant site- or region-specific basis. Each impact should be analyzed in proportion to its significance.

   Data provided in the applicant's ER are adequate if they describe:

   - the degree to which the local environmental resources would be affected by use of replacement power alternatives. These data are in agreement with data obtained from other sources, when available.

   - the significance or potential significance of such environmental impacts. SMALL impacts result when no discernible change in environmental resources occurs as a result of using replacement power alternatives. MODERATE impacts result when there is a discernible change. LARGE impacts occur when there is substantial disruption of environmental resources.

- any mitigative measures for which credit is being taken to reduce environmental concerns

Supplemental data obtained from other individuals and organizations may be useful in determining the completeness of the applicant's identification of housing impacts.

5. Consider and evaluate potential mitigation measures or alternatives that might reduce or eliminate the adverse impacts or the disproportionate distribution of the impacts in those cases where the impacts are MODERATE or LARGE. These may have been considered in the applicant's ER.

6. Based on the results of the assessments listed above, prepare the following for the SEIS:

- a summary statement (qualitative or quantitative, as appropriate) about the degree to which environmental resources are expected to receive impacts from replacement power alternatives, together with the significance of these impacts

- a discussion of the reasoning (e.g., based on locations and changes in population, local government revenue base, ecological impacts on other nearby plant sites or transmission corridors) behind the estimated degree of impact

- a discussion of any mitigative measures for which credit is being taken to reduce environmental concerns

IV. EVALUATION FINDINGS

The depth and extent of the information in the SEIS would be governed by the extent and significance of the effects of replacement power alternative fuel cycles. The reviewer should verify that sufficient information is available to meet the relevant requirements.

V. IMPLEMENTATION

The method described in this ESRP would be used by the staff in evaluating conformance with the Commission's regulations, except in those cases in which the applicant for license renewal proposes an acceptable alternative for complying with specified portions of the regulations.

VI. BIBLIOGRAPHY

10 CFR Part 51. *Code of Federal Regulations*, Title 10, *Energy,* Part 51, "Environmental Protection Regulations for Domestic Licensing and Related Regulatory Functions."

U.S. Nuclear Regulatory Commission (NRC). 2013. *Generic Environmental Impact Statement* (GEIS) *for License Renewal of Nuclear Plants.* NUREG-1437, Vols. 1, 2, and 3, Revision 1. Washington, D.C.

U.S. NUCLEAR REGULATORY COMMISSION

ENVIRONMENTAL STANDARD
REVIEW PLAN

OFFICE OF NUCLEAR REACTOR REGULATION

4.11.3 ENVIRONMENTAL CONSEQUENCES FROM THE TERMINATION OF NUCLEAR PLANT OPERATIONS AND DECOMMISSIONING

I. AREAS OF REVIEW

This environmental standard review plan (ESRP) provides guidance for the preparation of environmental changes and impacts from the termination of nuclear plant operations and decommissioning and preparation of input to the supplemental environmental impact statement (SEIS).

The scope includes (1) review of the discussion of the termination of nuclear plant operations and decommissioning and the potential changes in the environmental impacts resulting from continued operation, (2) identification and evaluation of new information (if any) related to potential changes in environmental impacts from the termination of plant operations and decommissioning resulting from continued operation during the renewal term for significance, and (3) preparation of input to the SEIS presenting the analyses related to those Category 1 issues.

Data and Information Needs

The types of data and information needed would be affected by nuclear power plant site- and plant-specific factors. The following data or information may be needed:

*   a description of the applicant's process for identifying new and potentially significant information on environmental issues related to changes in decommissioning impacts resulting from continued operation during the renewal term

---

- new information on the environmental issues related to changes in decommissioning impacts resulting from continued operation during the renewal term known to the applicant

- new and potentially significant information on the changes in environmental impacts of decommissioning resulting from operation during the renewal term identified by the public

II. ACCEPTANCE CRITERIA

Acceptance criteria for the evaluation of changes in environmental impacts of decommissioning resulting from continued operation during the renewal term are based on the relevant requirements of the following regulations:

- 10 CFR 51.53(c), concerning the content of environmental reports (ERs) and required analyses

- 10 CFR 51.70(b), concerning independent evaluation of the assessment and the reliability of information used in the assessment

- 10 CFR 51.71, concerning compliance with environmental-quality standards and requirements that have been imposed by Federal, State, regional, local, and affected American Indian Tribal agencies

- 10 CFR 51.95(c), concerning consideration of significant new information

- 10 CFR 51, Subpart A, Appendix B, concerning Commission's findings on the scope and magnitude of environmental impacts renewing the operating license for a nuclear power plant

Technical Rationale

The technical rationale for evaluating potential changes in the environmental impacts from the termination of plant operations and decommissioning resulting from continued operation is discussed in the following paragraph:

The NRC staff is required by 10 CFR 51.95(c)(4) to integrate conclusions, as amplified by the supporting information in the GEIS, for issues designated as Category 1 and any significant new information in an environmental impact statement (EIS) prepared at the license renewal stage. The review conducted under this ESRP leads to preparation of sections related to changes in environmental impacts from the termination of plant operations and decommissioning resulting from continued operation during the renewal term, as appropriate, and address significant new information, if any.

III. REVIEW PROCEDURES

Suggested steps for the review process are as follows:

1. Review the discussion of the issue in the GEIS to identify the information considered and the conclusions reached. The step establishes the base for evaluation of information identified by the applicant, the public, and the staff. The following table lists the termination of plant operations and decommissioning issues that were addressed in the GEIS for which generic conclusions were reached:

Table B-1. Summary of Findings on NEPA Issues for License Renewal of Nuclear Power Plants		
Issue	Category	Finding
Termination of Nuclear Power Plant Operations and Decommissioning		
Termination of plant operations and decommissioning	1	SMALL. License renewal is expected to have a negligible effect on the impacts of terminating operations and decommissioning on all resources.

2. Determine if there is new information on this issue that should be evaluated. The following sources of information should be included in the search for new information:

   - the applicant's ER. When reviewing, consider the applicant's process for discovering new information and evaluating the significance of any new information discovered

   - records of public meetings and correspondence related to the application

   - environmental quality standards and regulations

   - If the search conducted in this step reveals new information, then continue with Step 3

3. Evaluate the significance of new information. If new information is significant, prepare a concise statement(s) of the issues raised by significant new information and provide these statements to the Project Manager (PM) for review and disposition.

IV. EVALUATION FINDINGS

The depth and extent of the input to the SEIS would be governed by the extent of the analysis required to reach conclusions on issues related to the potential changes to environmental impacts from the termination of plant operations and decommissioning resulting from continued operation during the renewal term. The information that should be included in the SEIS is described in the review procedures.

V. IMPLEMENTATION

The method described in this ESRP would be used by the staff in evaluating conformance with the Commission's regulations, except in those cases in which the applicant for license renewal proposes an acceptable alternative for complying with specified portions of the regulations.

VI. BIBLIOGRAPHY

10 CFR Part 51. *Code of Federal Regulations*, Title 10, *Energy,* Part 51, "Environmental Protection Regulations for Domestic Licensing and Related Regulatory Functions."

National Environmental Policy Act of 1969 (NEPA). (42 USC 4321 et seq.).

U.S. Nuclear Regulatory Commission (NRC). 2013. *Generic Environmental Impact Statement for License Renewal of Nuclear Plants.* NUREG-1437, Vols. 1, 2, and 3, Revision 1. Office of Nuclear Reactor Regulation, Washington, D.C.

NUREG-1555, Supplement 1

U.S. NUCLEAR REGULATORY COMMISSION
ENVIRONMENTAL STANDARD REVIEW PLAN
OFFICE OF NUCLEAR REACTOR REGULATION

4.12 CUMULATIVE IMPACTS

I. AREAS OF REVIEW

This environmental standard review plan (ESRP) provides guidance for the analysis and assessment of cumulative impacts. Issues assessed here were identified as plant-specific (Category 2) in the *Generic Environmental Impact Statement for License Renewal of Nuclear Plants* (GEIS; NUREG-1437, Volumes 1, 2, and 3, Revision 1) and in Table B-1 of Appendix B, Subpart A to 10 CFR 51.

The scope for each individual section includes (1) review of the cumulative impacts issue in the GEIS, (2) evaluation of the data and analysis in the applicant's ER, (3) analysis and evaluation of the data, if appropriate, and (4) preparation of input to the supplemental environmental impact statement (SEIS).

II. ACCEPTANCE CRITERIA

Cumulative impacts is a Category 2 issue and requires a plant-specific analysis (see 10 CFR 51.53(c)(3)(ii)(O)). A cumulative impact is defined by the Council on Environmental Quality (CEQ) in in 40 CFR 1508.7 as an "impact on the environment which results from the incremental impact of the action when added to other past, present, and reasonably foreseeable future actions regardless of what agency (Federal or non-Federal) or person undertakes such other actions." Actions to be considered in cumulative impact analyses include new and continuing activities, such as activities associated with license renewal (e.g., continued reactor operations and refurbishment), that are conducted, regulated, or approved by a Federal agency. The cumulative impacts analysis takes into account all actions, however minor, since impacts from individually minor actions may be significant when considered collectively over time. The goal of the analysis is to identify potentially significant impacts to improve decisions and move toward more sustainable development (CEQ 1997).

4.12-1

III. REVIEW PROCEDURES

Suggested steps for the review process are as follows:

1. Review the discussion of cumulative impacts in the GEIS to identify the information considered and the conclusions reached. This step establishes the base for evaluation of information identified by the applicant, the public, and the staff. The Category 2 issue identified in the GEIS is listed in the following table.

Table B-1. Summary of Findings on NEPA Issues for License Renewal of Nuclear Power Plants		
Cumulative Impacts		
Issue	**Category**	**Finding**
Cumulative impacts	2	Cumulative impacts of continued operations and refurbishment associated with license renewal must be considered on a plant-specific basis. Impacts would depend on regional resource characteristics, the resource-specific impacts of license renewal, and the cumulative significance of other factors affecting the resource.

2. The analysis of cumulative impacts should focus on the resources that could be affected by the incremental impacts of continued plant operations. These resource areas include:

   - Air Quality and Noise
   - Geology and Soils
   - Water Resources
   - Ecological Resources
   - Historic and Cultural Resources
   - Socioeconomics
   - Human Health
   - Environmental Justice
   - Waste Management
   - Global Climate Change

3. For each resource area, establish the following:

   - The geographic scope (i.e., regions of influence). The regions of influence encompass the areas of affect and the distances at which impacts associated with license renewal may occur. Geographic boundaries may vary by the resource area being evaluated and the distances over which an impact may occur (e.g., the evaluation of impacts on air quality may have a greater regional extent than that of impacts on cultural resources).

   - The time frame for the analysis. The time frame incorporates the sum of the effects of renewal in combination with past, present, and future actions, since impacts may accumulate or develop over time. The reasonably foreseeable time frame for future actions evaluated is 20 years (based on the typical license renewal term) from the time the license renewal is granted.

- The potential impacting factors of each past, present, or reasonably foreseeable future action or activity. Both the license renewal and other actions (related and nonrelated, including trends such as global climate change) would generate factors that could contribute to cumulative impacts. The impacts of activities associated with the proposed action (license renewal) should be discussed for each resource area listed above.

4. Identify resource areas (e.g., water and aquatic resources) where the contributions of ongoing actions within a region are regulated and monitored through a permitting process (e.g., NPDES) under State or Federal authority. In these cases, it may be assumed that cumulative impacts are managed as long as these actions (facilities) are in compliance with their respective permits.

5. Analyze cumulative environmental impacts, as follows:

- Identify and calculate the likely cumulative environmental impacts for the area(s) or region(s) currently or likely to be affected by power plant operations and/or refurbishment activities associated with license renewal.

- Describe the impacts in sufficient detail so that reviewers may compare the cumulative impacts with those of renewing the operating license. The impacts analyses should include direct and indirect effects . A direct effect is an effect that is caused by the action and occurs at the same time and place. An indirect effect is caused by the action and is later in time or farther removed in distance, but still reasonably foreseeable. Impacts should be analyzed in proportion to its significance.

Data provided in the applicant's ER are adequate if they describe:

- the degree to which the local environmental resources would be affected. These data are in agreement with data obtained from other sources, when available.

- the significance or potential significance of such cumulative environmental impacts.

- any mitigative measures for which credit is being taken to reduce environmental concerns

Supplemental data obtained from other individuals and organizations may be useful in determining the completeness of the applicant's identification of cumulative impacts.

6. Based on the results of the assessments listed above, prepare the following for the SEIS:

- a summary statement (qualitative or quantitative, as appropriate) about the degree to which environmental resources are expected to experience cumulative impacts

- a discussion of the reasoning (e.g., based on locations and changes in population, local government revenue base, ecological impacts on other nearby plant sites) behind the estimated degree of cumulative impact

- a discussion of any mitigative measures for which credit is being taken to reduce environmental concerns

IV. EVALUATION FINDINGS

The depth and extent of the input to the SEIS would be governed by the extent of the analysis required to reach conclusions on potential cumulative impact caused by the added contribution from continued operations during the license renewal term and refurbishment impacts associated with license renewal. The information that should be included in the SEIS is described in the review procedures.

V. IMPLEMENTATION

The method described in this ESRP would be used by the staff in evaluating conformance with the Commission's regulations, except in those cases in which the applicant for license renewal proposes an acceptable alternative for complying with specified portions of the regulations.

VI. BIBLIOGRAPHY

10 CFR Part 51. *Code of Federal Regulations*, Title 10, *Energy,* Part 51, "Environmental Protection Regulations for Domestic Licensing and Related Regulatory Functions."

Council on Environmental Quality (CEQ). 1997. *Considering Cumulative Effects under the National Environmental Policy Act.* Washington, D.C. January.

National Environmental Policy Act of 1969 (NEPA). 42 USC 4321 et seq.

U.S. Environmental Protection Agency (EPA). 1999. Consideration of Cumulative Impacts in EPA Review of NEPA Documents, EPA 315-R-99-002, Washington, D.C. May.

U.S. Nuclear Regulatory Commission (NRC). 2013. *Generic Environmental Impact Statement for License Renewal of Nuclear Plants.* NUREG-1437, Vols. 1, 2, and 3, Revision 1. Office of Nuclear Reactor Regulation, Washington, D.C.

NUREG-1555, Supplement 1

U.S. NUCLEAR REGULATORY COMMISSION

ENVIRONMENTAL STANDARD REVIEW PLAN

OFFICE OF NUCLEAR REACTOR REGULATION

4.13 REFERENCES

I. AREAS OF REVIEW

This environmental standard review plan (ESRP) provides guidance for the listing of references cited in Chapter 4 of the supplemental environmental impact statement (SEIS).

II. ACCEPTANCE CRITERIA

Acceptance criteria for the preparation of the reference list are based on the relevant requirements of the following regulation:

- 10 CFR 51.70(b), concerning preparation of a draft EIS that is concise, clear, analytical, and written in plain language

III. REVIEW PROCEDURES

The environmental project manager (EPM) should contact reviewers for ESRPs 4.1 through 4.12 and compile a list of references cited in the SEIS sections that the reviewers have prepared. The citations should be checked for completeness and accuracy and prepared for inclusion in the SEIS.

4.13-1

IV. EVALUATION FINDINGS

The reviewer of information covered by this ESRP should prepare the SEIS section that lists references cited in the SEIS sections covering environmental impacts. The completed reference list constitutes the findings for this ESRP.

V. IMPLEMENTATION

The method described in this ESRP would be used by the staff in evaluating conformance with the Commission's regulations, except in those cases in which the applicant for license renewal proposes an acceptable alternative for complying with specified portions of the regulations.

VI. BIBLIOGRAPHY

10 CFR Part 51. *Code of Federal Regulations*, Title 10, *Energy*, Part 51, "Environmental Protection Regulations for Domestic Licensing and Related Regulatory Functions."

U.S. Nuclear Regulatory Commission. 2013. *Generic Environmental Impact Statement for License Renewal of Nuclear Plants*. NUREG-1437, Vols. 1, 2, and 3, Revision 1. Office of Nuclear Reactor Regulation, Washington, D.C.

U.S. NUCLEAR REGULATORY COMMISSION
ENVIRONMENTAL STANDARD
REVIEW PLAN
OFFICE OF NUCLEAR REACTOR REGULATION

5.0 ENVIRONMENTAL IMPACTS OF POSTULATED ACCIDENTS

I. AREAS OF REVIEW

This environmental standard review plan (ESRP) provides guidance for the preparation of introductory paragraphs for the portion of the supplemental environmental impact statement (SEIS) that describes environmental impacts of postulated plant accidents during the license renewal term.

The scope of this plan is the development of paragraphs that introduce the material from the reviews conducted under ESRPs 5.1 and 5.2. It includes the description of the environmental issues associated with postulated accidents discussed in the *Generic Environmental Impact Statement for License Renewal of Nuclear Plants* (GEIS; NUREG-1437, Volumes 1, 2, and 3, Revision 1).

II. ACCEPTANCE CRITERIA

The reviewer should ensure that the introductory paragraphs prepared under this ESRP are consistent with the intent of the following regulations:

- 10 CFR 51.53(c), concerning the content of environmental reports (ERs) and required analyses

- 10 CFR 51.70(b), concerning preparation of a draft environmental impact statement (DEIS) that is concise, clear, analytic, and written in plain language

- 10 CFR 51.71, concerning preparation of a DEIS at the license renewal stage

- 10 CFR 51.95(c), concerning preparation of a final environmental impact statement (EIS) at the license renewal stage

- 10 CFR 51, Subpart A, Appendix B, Table B-1, concerning findings on environmental issues for license renewal of nuclear power plants

Technical Rationale

The technical rationale for evaluating the applicant's description of the potential environmental impacts of postulated accidents during the renewal term is discussed in the following paragraph:

The NRC staff is required by 10 CFR 51.95(c)(4) to integrate conclusions, as amplified by the supporting information in the GEIS, for issues that are designated as Category 1 or resolved Category 2, information developed for those open Category 2 issues applicable to the plant, and any significant new information in an EIS prepared at the license renewal stage. The review conducted under this ESRP leads to preparation of introductory paragraphs that orient the reader concerning the relevance of the material to the overall organization and goals of the SEIS and add clarity to the presentation.

III. REVIEW PROCEDURES

The material to be prepared is informational in nature, and no specific analysis of data is required. The environmental issue associated with operation during the renewal term that was considered in the GEIS and determined to be a Category 1 issue is listed in the following table.

Table B-1. Summary of Findings on NEPA Issues for License Renewal of Nuclear Power Plants		
Issue	Category	Finding
Postulated Accidents		
Design-basis accidents	1	SMALL. The NRC staff has concluded that the environmental impacts of design-basis accidents are of small significance for all plants.

Generic conclusions relative to impacts were reached for those issues that are appropriate for all plants, or for some issues for specific classes of plants. These conclusions were that (1) a single level of significance could be assigned to the impact and (2) plant-specific mitigation measures are not likely to be sufficiently beneficial to warrant implementation. The generic analysis of severe accidents analysis described in the GEIS applies to all plants. It concludes that the probability-weighted consequences of atmospheric releases, fallout onto open bodies of water, releases to groundwater, and societal and economic impacts of severe accidents are of small significance. In the absence of new and significant information, these issues may be addressed in the SEIS without additional plant-specific analysis.

Environmental issues considered in the GEIS for which these conclusions could not be reached for all plants, or for specific classes of plants, are Category 2 issues. The Category 2 issue related to postulated accidents is listed in the following table.

Table B-1. Summary of Findings on NEPA Issues for License Renewal of Nuclear Power Plants		
Issue	Category	Finding
Postulated Accidents		
Severe accidents	2	SMALL. The probability-weighted consequences of atmospheric releases, fallout onto open bodies of water, releases to groundwater, and societal and economic impacts from severe accidents are small for all plants. However, alternatives to mitigate severe accidents must be considered for all plants that have not considered such alternatives.

Not all plants have performed analyses of the measures that could mitigate the consequences of severe accidents. Consequently, alternatives to mitigate severe accidents are a Category 2 issue for plants that have not performed a plant-specific evaluation of severe accident mitigation alternatives and submitted that evaluation to the Commission for review. A plant-specific analysis is required for this issue.

If there is new and significant information related to the environmental impacts associated with postulated accidents identified by the applicant, members of the public, or the staff during the environmental review, the reviewer for this ESRP should prepare a table that directs readers to the SEIS sections dealing with the issues.

IV. EVALUATION FINDINGS

The reviewer of information covered by this ESRP should prepare introductory paragraphs for the SEIS. The paragraph(s) should introduce the nature of the material to be presented by the reviewers of information covered by ESRPs 5.1 and 5.2. The paragraph(s) should list the types of information to be presented and describe their relationships to information presented earlier and to be presented later in the SEIS.

V. IMPLEMENTATION

The method described in this ESRP would be used by the staff in evaluating conformance with the Commission's regulations, except in those cases in which the applicant for license renewal proposes an acceptable alternative for complying with specified portions of the regulations.

VI. BIBLIOGRAPHY

10 CFR Part 51. *Code of Federal Regulations*, Title 10, *Energy*, Part 51, "Environmental Protection Regulations for Domestic Licensing and Related Regulatory Functions."

U.S. Nuclear Regulatory Commission. 2013. *Generic Environmental Impact Statement for License Renewal of Nuclear Plants*. NUREG-1437, Vols. 1, 2, and 3, Revision 1. Office of Nuclear Reactor Regulation, Washington, D.C.

U.S. NUCLEAR REGULATORY COMMISSION

ENVIRONMENTAL STANDARD REVIEW PLAN

OFFICE OF NUCLEAR REACTOR REGULATION

5.1 POSTULATED PLANT ACCIDENTS

I. AREAS OF REVIEW

This environmental standard review plan (ESRP) provides guidance for the review of environmental impacts of postulated plant accidents during the license renewal term and preparation of input to the supplemental environmental impact statement (SEIS). These issues are discussed in Section 4.9.1.2 of the *Generic Environmental Impact Statement for License Renewal of Nuclear Plants* (GEIS; NUREG-1437, Volumes 1, 2, and 3, Revision 1).

The scope includes (1) review of the GEIS discussion of postulated accidents, (2) identification and evaluation of new information related to environmental impacts of postulated accidents during the renewal term for significance, (3) preparation of input to the SEIS that dispositions the Category 1 issue, and (4) preparation of input to the SEIS that introduces the discussion of the Category 2 issue.

Impacts of design-basis accidents (DBAs) during the renewal term is a Category 1 issue, and the impacts of severe accidents is a Category 2 issue. The probability-weighted consequences of atmospheric releases to groundwater and societal and economic impacts from severe accidents are small for all plants. However, alternatives to mitigate severe accidents must be considered for all plants that have not considered such alternatives. A plant-specific review of alternatives to mitigate severe accidents is required by 10 CFR 51.53(c)(3)(ii)(L) if the NRC staff has not previously considered severe accident mitigation alternatives for the applicant's plant in an environmental impact statement (EIS), SEIS, or environmental assessment. If a SAMA review has been conducted, then only new and significant information should be evaluated in accordance with ESRP 5.2.

<u>Data and Information Needs</u>

The types of data and information needed would be affected by nuclear power plant site- and plant-specific factors; the level of detail should be scaled according to the anticipated magnitude of the potential impacts. The following data or information may be needed:

- a description of the applicant's process for identifying new and potentially significant information on environmental issues related to postulated accidents during the renewal term

- new information on environmental impacts of postulated plant accidents during the renewal term known to the applicant

- new and potentially significant information on environmental impacts of postulated plant accidents during the license renewal term identified by the public

II. ACCEPTANCE CRITERIA

Acceptance criteria for the evaluation of postulated plant accidents during the renewal term are based on the relevant requirements of the following regulations:

- 10 CFR 51.45(b), concerning environmental considerations in the applicant's ER

- 10 CFR 51.45(d), concerning discussion of compliance with applicable environmental quality standards and requirements in the applicant's ER

- 10 CFR 51.53(c)(3)(ii), concerning analyses required in ERs submitted at the license renewal stage

- 10 CFR 51.70(b), concerning an independent evaluation of the assessment and the reliability of information used in the assessment

- 10 CFR 51.71(d), concerning content requirements that apply to analyses in draft environmental impact statements at the license renewal stage

- 10 CFR 51.95(c)(4), concerning contents of SEIS and consideration of significant new information

- 10 CFR 51 Subpart A, Appendix B, concerning findings on environmental issues for license renewal of nuclear power plants.

<u>Technical Rationale</u>

The technical rationale for evaluating the applicant's description of postulated plant accidents during the renewal term is discussed in the following paragraph:

The NRC staff is required by 10 CFR 51.95(c)(4) to integrate conclusions, as amplified by the supporting information in the GEIS, for issues designated as Category 1 or resolved Category 2, information developed for those open Category 2 issues applicable to the plant, and any significant new information in an EIS prepared at the license renewal stage. The review conducted under this ESRP leads to preparation of sections of the SEIS that incorporate the conclusions from the GEIS related to postulated plant accidents during the renewal term, as appropriate, and address significant new information, if any.

III. REVIEW PROCEDURES

Suggested steps for the review process are as follows:

1. Review the discussion of the issue in the GEIS to identify the information considered and the conclusions reached. This step establishes the base for evaluation of information identified by the applicant, the public, and the staff. The following table lists the postulated plant accidents issue that was addressed in the GEIS for which generic conclusions were reached.

Table B-1. Summary of Findings on NEPA Issues for License Renewal of Nuclear Power Plants		
Issue	Category	Finding
Postulated Accidents		
Design-basis accidents	1	SMALL. The NRC staff has concluded that the environmental impacts of design-basis accidents are of small significance for all plants.
Severe accidents	2	SMALL. The probability-weighted consequences of atmospheric releases, fallout onto open bodies of water, releases to groundwater, and societal and economic impacts from severe accidents are small for all plants. However, alternatives to mitigate severe accidents must be considered for all plants that have not considered such alternatives.

2. Determine if there is new information on this issue that should be evaluated. The following sources of information should be included in the search for new information:

   • The applicant's ER. An applicant is required by 10 CFR 51.53(c)(3)(iv) to disclose new and significant information of environmental impacts of license renewal of which it is aware. In reviewing the applicant's ER, consider the applicant's process for discovering new information related to environmental impacts of postulated accidents and evaluating the significance of any new information discovered.

   • Records of public meetings and correspondence related to the application. Compare information presented by the public with information considered in the GEIS.

   • Environmental standards and regulations. Have the applicable environmental quality standards and regulations changed since the analysis leading to the GEIS? If so, do the changes affect the NRC evaluation of applications for license renewal?

If the search conducted in this step reveals new information, then continue with Step 3. Otherwise, prepare the section for SEIS describing the search for new information, stating the conclusion that there is none, and adopting conclusions from the GEIS.

3. Evaluate the significance of new information.

4. Prepare the section for the SEIS describing the search for new information, summarizing new information found, and presenting results of evaluation of significance.

IV. EVALUATION FINDINGS

The depth and extent of the input to the SEIS would be governed by the extent of the analysis required to reach a conclusion related to the environmental impacts of postulated accidents during the renewal term. The information that should be included in the SEIS is described in the review procedures. In accordance with the Commission's direction in the SRM for SECY-12-0063—Final Rule, when reiterating the conclusion of the GEIS in the evaluation findings, the following entire phrase shall be included in the text: "…the probability-weighted consequences of severe accidents are SMALL."

V. IMPLEMENTATION

The method described in this ESRP would be used by the staff in evaluating conformance with the Commission's regulations, except in those cases in which the applicant for license renewal proposes an acceptable alternative for complying with specified portions of the regulations.

VI. BIBLIOGRAPHY

10 CFR Part 50, "Domestic Licensing of Production and Utilization Facilities."

10 CFR 51.45, "Environmental report."

10 CFR 51.53, "Postconstruction environmental reports."

10 CFR 51.70, "Draft environmental impact statement general."

10 CFR 51.71, "Draft environmental impact statement contents."

10 CFR 51.95, "Postconstruction environmental impact statements."

10 CFR 51 Subpart A, Appendix B, "Environmental Effect of Renewing the Operating License of a Nuclear Power Plant."

U.S. Nuclear Regulatory Commission. 2012. Staff Requirements, SRM-SECY-12-0063 – Final Rule: Revisions to Environmental Review for Renewal of Nuclear Power Plant Operating Licenses (10 CFR Part 51; RIN 3150–AI42). December 6, 2012. ADAMS Accession No. M121206A.

U.S. NUCLEAR REGULATORY COMMISSION
ENVIRONMENTAL STANDARD
REVIEW PLAN
OFFICE OF NUCLEAR REACTOR REGULATION

5.2 SEVERE ACCIDENT MITIGATION ALTERNATIVES

I. AREAS OF REVIEW

This environmental standard review plan (ESRP) provides guidance for the analysis and assessment of the severe accident mitigation alternatives (SAMAs). This issue was identified as a Category 2 issue in the *Generic Environmental Statement for License Renewal of Nuclear Plants* (GEIS; NUREG-1437, Volumes 1, 2, and 3, Revision 1), and in Table B-1 of Appendix B, Subpart A to 10 CFR 51. An applicant for license renewal is required by 10 CFR 51.53(c)(3)(ii)(L) to consider alternatives to mitigate severe accidents at the plant if the staff has not previously considered severe accident mitigation alternatives for the applicant's plant in an environmental impact statement (EIS) or related supplement or in an environmental assessment for the plant.

The scope includes an analysis of SAMAs and the preparation of an appropriate statement for the supplemental environmental impact statement (SEIS). The analysis of SAMAs includes the identification and evaluation of alternatives that reduce the radiological risk from a severe accident by preventing substantial core damage (i.e., preventing a severe accident) or by limiting releases from containment in the event that substantial core damage occurs (i.e., mitigating the impacts of a severe accident). The purpose of the review is to ensure that plant and procedure changes with the potential for improved severe accident safety performance are identified and evaluated.

Data and Information Needs

The type of data and information needed would be affected by nuclear power plant site- and plant-specific factors. The following data or information should be obtained:

- a list of leading contributors to (1) core-damage frequency (e.g., from dominant severe accident sequences or initiating events), and (2) dose consequences (e.g., from each release class and associated source term) (from the environmental report [ER])

- the methodology, process, and rationale used by the applicant to identify, screen, and select alternatives (from the ER)

- the estimated cost, risk reduction, and dollar benefits for the selected SAMAs and the assumptions used to make these estimates (from the ER)

- a description and list of any alternatives that have been or would be implemented to reduce the risk of a severe accident, or that would be further evaluated by the applicant for possible future implementation (from the ER)

II. ACCEPTANCE CRITERIA

Acceptance criteria for the analysis and evaluation of SAMAs are based on the relevant requirements of the following regulations:

- 10 CFR 51.53(c)(3)(ii)(L), concerning the need to consider alternatives to mitigate severe accidents for the applicant's plant if SAMAs were not previously considered

- 10 CFR 51.70(b), concerning permitting an independent evaluation of the assessment and the reliability of information used in the assessment

- 10 CFR 51, Subpart A, Appendix B, Table B-1, concerning findings on environmental issues for license renewal of nuclear power plants

Additional regulatory positions and specific criteria in support of the regulations identified above are as follows:

- Interim Policy Statement, "Power Plants—Nuclear Power Plant Accident Considerations under NEPA" (NRC 1980), provides guidance concerning the early consideration of either additional features or other actions that would prevent or mitigate the consequences of serious accidents.

- "Policy Statement on Severe Reactor Accidents Regarding Future Designs and Existing Plants" (NRC 1985) provides guidance for dealing with severe accident issues and vulnerabilities for operating reactors.

- Policy Statement, "Safety Goals for the Operations of Nuclear Power Plants" (NRC 1986) provides safety goals for nuclear power plants.

- NUREG/BR-0184, *Regulatory Analysis Technical Evaluation Handbook* (NRC 1997b), provides guidance concerning the value impact methodology.

- NUREG/BR-0058, Rev. 4, *Regulatory Analysis Guidelines of the U.S. Nuclear Regulatory Commission* (NRC 2004) states the policy for the preparation and the contents of regulatory analyses, including estimation of values and impacts for alternatives and the "dollars per person-rem" conversion factors.

- NUREG/CR-6349 (Mubayi et al. 1995) provides information on dollars per person-rem conversion factor for offsite damage costs.

- Generic Letter 88-20 (NRC 1988) provides guidance on the performance of an individual plant examination at operating plants for severe accident vulnerabilities.

- Generic Letter 88-20, Supplement 3 (NRC 1990a) provides guidance on accident prevention and mitigation features identified in the Containment Performance Improvement Program that may be valid for consideration in the review of SAMAs.

- Generic Letter 88-20, Supplement 4 (NRC 1991) provides guidance on conducting an individual plant examination for externally initiated events.

- Regulatory Guide 4.2, Supplement 1, Revision 1, *Preparation of Environmental Reports for Nuclear Power Plant License Renewal Applications* (NRC 2013a) provides guidance on preparation of ERs associated with license renewal.

- Regulatory Guides 1.174 (NRC 2002b) and 1.200 (NRC 2007) provide guidance on general concepts in use and evaluation of probabilistic risk assessments for risk-informed decisions.

In addition to the above, the reviewer should be familiar with NEI 05-01, "SAMA Analysis Guidance Document," which is the nuclear industry's guidance document describing how to perform the SAMA analysis and describes the information that should be included in the SAMA analysis portion of the ER.

The following acceptance criterion is used:

- Completeness and reasonableness concerning (1) the identification of SAMAs applicable to the plant under consideration, (2) the estimation of core damage frequency reduction and averted person-rem for each SAMA, (3) the estimation of cost for each SAMA, (4) the screening criteria to identify SAMAs for further consideration, and (5) the final disposition of promising SAMAs.

Technical Rationale

The technical rationale for evaluating the applicant's SAMAs is discussed in the following paragraphs:

The SEIS should include an analysis of the SAMAs for the applicant's plant if they have not previously been considered in an EIS or related supplement or in an environmental assessment. The purpose is to review and evaluate plant design alternatives and procedural changes that could significantly reduce the radiological risk from a severe accident by preventing substantial core damage (i.e., preventing a severe accident) or by limiting releases from containment in the event that substantial core damage occurs (i.e., mitigating the impacts of a severe accident).

In 1980, the NRC published an interim policy statement (Interim Policy Statement, "Nuclear Power Plant Accident Considerations Under the National Environmental Policy Act of 1969" [NRC 1980]) that stated that it was the intent of the Commission for the staff to take steps to identify additional cases that might warrant early consideration of either additional features or other actions that would prevent or mitigate the consequences of serious accidents.

In 1985, the NRC published a policy statement ("Policy Statement on Severe Reactor Accidents Regarding Future Designs and Existing Plants," August 9, 1985 [NRC 1985]). It concluded that existing plants posed no undue risk to public health and safety and that there is no present basis for immediate action on generic rulemaking or other regulatory changes for these plants because of severe-accident risk. However, the policy statement indicated that "the Commission plans to formulate an approach for a systematic safety examination of existing plants to determine whether particular accident vulnerabilities are present and what cost-effective changes are desirable to ensure that there is no undue risk to public health and safety."

A 1989 court decision (*Limerick Ecology Action vs. NRC*, 869 F.2d 719 [3rd Cir. 1989]) stated that the "Action of NRC in addressing SAMDAs (prior term for SAMAs) through policy statements, not rule making, did not satisfy NEPA, where policy statements did not represent requisite careful consideration of environmental consequences, excluded consideration of design alternatives without making any conclusions about effectiveness of any particular alternative, and issues were not generic in that impact of SAMDAs on environment would differ with a particular plant's design, construction and locations." NRC considers the evaluation of SAMAs in the environmental impact review that is performed as part of every application for a license renewal if SAMAs have not been considered for the plant.

III. REVIEW PROCEDURES

Suggested steps for conducting the review are as follows:

1. Determine if the staff previously considered SAMAs for the applicant's plant in an EIS or related supplement or in an environmental assessment. If not, then continue the analysis at Step 2. Otherwise, prepare a statement for the SEIS that describes the SAMA analysis and identifies the location of the analysis.

2. Become familiar with analyses, the process, and design alternatives considered in previous studies, including the following:

    - Limerick (NRC 1989a)

    - NRC Containment Performance Improvement Program (NRC 1989b and 1990b)

    - Comanche Peak (NRC 1989c)

    - Watts Bar (NRC 1995)

    - System 80+ (NRC 1997c)

    - ABWR (NRC 1997d)

    - GEIS for License Renewal and plant-specific supplements

3. Evaluate the applicant's methods for establishing the risk profile for the plant, including the core damage frequency, population dose risk, and offsite economic cost risk for both internal and external events. This review should consider the evolution and peer reviews of the Level 1 (core damage frequency) and Level 2 (containment performance) probabilistic risk assessment models and the process used to extend the analysis to an assessment of offsite consequences. The scope of the review includes the major inputs to the MACCS2 offsite consequence code (NRC 1998), including plant-specific inputs related to core inventory, meteorology, population, evacuation, and economic impacts.

4. Evaluate the applicant's methods for identifying the potential mitigation alternatives. If the applicant used an alternative methodology to a probabilistic risk assessment approach to assess potential SAMAs, the staff evaluation should be appropriately modified. For example, seismic margins analysis does not produce a CDF (i.e., it is a qualitative analysis) and is predicated on the ability to evaluate the seismic durability of equipment required to safely shut the plant down. The results of this kind of analysis do not directly lend themselves to the frequency-based SAMA analysis. Alternative cost-benefit approaches are appropriate when a margins method has been used to screen external events.

    - Determine if this set of potential alternatives represents a reasonable range of preventive and mitigation alternatives.

    - Verify that the applicant's list of potential SAMAs includes a reasonable range of applicable SAMAs derived from consideration of previous analyses and based on insights from the Level 1 and Level 2 portions of the applicant's probabilistic risk assessment or individual plant examination/individual plant examination of external events.

5. Evaluate the applicant's basis for estimating the degree to which various alternatives would reduce risk (expressed as a reduction in core-damage frequency or in terms of person-rem averted). NRC staff may make bounding assumptions or adjustments to the applicant's analysis to determine the magnitude of the potential risk reduction for each SAMA.

6. Evaluate whether the applicant's cost estimates for each SAMA are reasonable, and compare the cost estimates with estimates developed elsewhere (e.g., using previous SAMA evaluations or using accepted cost-estimation tools).

7. Evaluate the cost-benefit comparison to determine if it is consistent with the cost-benefit balance criteria and methodology given in NUREG/BR-0184 (NRC 1997b) and NUREG/BR-0058, Rev. 4, (NRC 2004), and further analyze any SAMAs with estimated implementation costs within a factor of 2 to 5 of the estimated dollar benefits to ensure that a sufficient margin is present to account for uncertainties in assumptions used to determine the cost and benefit estimates.

8. Subject any SAMAs that remain following the screening given above to further probabilistic and deterministic considerations, including a qualitative assessment of the following:

    •   the impact of additional benefits that could accrue for the SAMA if it would be effective in reducing risk from certain external events, as well as internal events

    •   the effects of improvements already made at the plant

    •   any operational disadvantage associated with the potential SAMA

9. Prepare a statement for the SEIS that describes the applicant's analysis and details the staff's review process. Any mitigation should be described along with the estimated costs and benefits. The risk reduction for the facility should be provided. The statement for the SEIS should identify and describe the mitigation measures considered and committed to by the applicant.

IV. EVALUATION FINDINGS

The depth and extent of the input to the SEIS would be governed by the extent of the analysis required to reach a conclusion related to the applicant's SAMA analysis. The information that should be included in the SEIS is described in the review procedures.

V. IMPLEMENTATION

The method described in this ESRP would be used by the staff in evaluating conformance with the Commission's regulations, except in those cases in which the applicant for license renewal proposes an acceptable alternative for complying with specified portions of the regulations.

VI. BIBLIOGRAPHY

10 CFR Part 51. *Code of Federal Regulations*, Title 10, *Energy,* Part 51, "Environmental Protection Regulations for Domestic Licensing and Related Regulatory Functions."

Limerick Ecology Action vs. NRC 869 F. 2D 719 [3rd Cir. 1989].

Mubayi, V., V. Sailor, and G. Anandalingam. 1995. *Cost-Benefit Considerations in Regulatory Analysis.* NUREG/CR-6349. U.S. Nuclear Regulatory Commission, Washington, D.C.

National Environmental Policy Act of 1969 (NEPA). 42 USC 4321 et seq.

Nuclear Energy Institute (NEI). 2005. *Severe Accident Mitigation Alternatives (SAMA) Analysis Guidance Document.* NEI 05-01, Rev. A. Washington, D.C.

U.S. Nuclear Regulatory Commission (NRC). 1980. *Nuclear Power Plant Accident Considerations Under the National Environmental Policy Act of 1969.* 45 FR 40101. Washington, D.C.

U.S. Nuclear Regulatory Commission (NRC). 1985. "Policy Statement on Severe Reactor Accidents Regarding Future Designs and Existing Plants." 50 FR 32138, Washington, D.C.

U.S. Nuclear Regulatory Commission (NRC). 1986. *Safety Goals for the Operations of Nuclear Power Plants: Policy Statement; Republication.* 51 FR 30028. Washington, D.C.

U.S. Nuclear Regulatory Commission (NRC). 1988. "Individual Plant Examination for Severe Accident Vulnerabilities." Generic Letter 88-20. Washington, D.C. November 23.

U.S. Nuclear Regulatory Commission (NRC). 1989a. Letter from U.S. NRC to G.A. Hunger, Jr. (Philadelphia Electric Company), "Subject: Supplement to the Final Environmental Statement—Limerick Generating Station, Units 1, 2, and 3. Supplement to NUREG-0974."

U.S. Nuclear Regulatory Commission (NRC). 1989b. *Mark I Containment Performance Improvement Program.* SECY-89-017. Washington, D.C.

U.S. Nuclear Regulatory Commission (NRC). 1990a. "Completion of Containment Performance Improvement Program and Forwarding Insights for Use in the Individual Plant Examination for Severe Accident Vulnerabilities." Generic Letter 88-20, Supplement 3, July 6, 1990, Washington, D.C.

U.S. Nuclear Regulatory Commission (NRC). 1990b. *Recommendations of Containment Performance Improvement Program for Plants with Mark II, Mark III, Ice Condenser, and Dry Containments.* SECY-90-120. Washington, D.C.

U.S. Nuclear Regulatory Commission (NRC). 1991. "Individual Plant Examination of External Events (IPEEE) for Severe Accident Vulnerabilities - 10 CFR 50.54(f)." Generic Letter 88-20, Supplement 4, June 28, 1991, Washington, D.C.

U.S. Nuclear Regulatory Commission (NRC). 1995. *Final Environmental Statement Related to the Operation of Watts Bar Nuclear Plant, Units 1 and 2.* NUREG-0498, Suppl. 1. Washington, D.C.

U.S. Nuclear Regulatory Commission (NRC). 1997a. *Individual Plant Examination Program: Perspectives on Reactor Safety and Plant Performance.* NUREG-1560. Washington, D.C.

U.S. Nuclear Regulatory Commission (NRC). 1997b. *Regulatory Analysis Technical Evaluation Handbook.* NUREG/BR-0184. Washington, D.C.

U.S. Nuclear Regulatory Commission (NRC). 1997c. *Final Environmental Assessment by the Office of Nuclear Reactor Regulation Relating to the Certification of the System 80+ Standard Nuclear Plant Design.* NUREG-1462. Washington, D.C.

U.S. Nuclear Regulatory Commission (NRC). 1997d. *Final Environmental Assessment by the Office of Nuclear Reactor Regulation Relating to the Certification of the U.S. Advanced Boiling Water Reactor Design.* NUREG-1503. Washington, D.C.

U.S. Nuclear Regulatory Commission (NRC). 1998. *Code Manual for MACCS2, User's Guide.* NUREG/CR-6613. Washington, D.C.

U.S. Nuclear Regulatory Commission (NRC). 2002a. *Perspectives Gained From the Individual Plant Examination of External Events (IPEEE) Program.* NUREG-1742. Washington, D.C.

U.S. Nuclear Regulatory Commission (NRC). 2002b. *An Approach for Using Probabilistic Risk Assessment in Risk-Informed Decisions on Plant-Specific Changes to the Licensing Basis.* Regulatory Guide 1.174, Rev. 1. Washington, D.C.

U.S. Nuclear Regulatory Commission (NRC). 2004. *Regulatory Analysis Guidelines of the U.S. Nuclear Regulatory Commission.* NUREG/BR-0058, Rev. 4. Washington, D.C.

U.S. Nuclear Regulatory Commission (NRC). 2007. *An Approach for Determining the Technical Adequacy of Probabilistic Risk Assessment Results for Risk-Informed Decisions on Plant-Specific Changes to the Licensing Basis.* Regulatory Guide 1.200, Rev. 1. Washington, D.C.

U.S. Nuclear Regulatory Commission (NRC). 2013a. *Preparation of Environmental Reports for Nuclear Power Plant License Renewal Applications.* Regulatory Guide 4.2, Supplement 1, Revision 1. Washington, D.C.

U.S. Nuclear Regulatory Commission (NRC). 2013b. *Generic Environmental Impact Statement for License Renewal of Nuclear Plants.* NUREG-1437, Vols. 1, 2, and 3, Revision 1. Washington, D.C.

5.3 REFERENCES

I. AREAS OF REVIEW

This environmental standard review plan (ESRP) provides guidance for listing references in this chapter of the supplemental environmental impact statement (SEIS).

II. ACCEPTANCE CRITERIA

Acceptance criteria for the preparation of the reference list are based on the relevant requirements of the following regulation:

- 10 CFR 51.70(b), concerning preparation of a draft environmental impact statement (EIS) that is concise, clear, analytic, and written in plain language.

III. REVIEW PROCEDURES

- The reviewer should contact reviewers for ESRPs 5.0 through 5.2 and compile a list of references cited in the SEIS sections that the reviewers have prepared. The citations should be checked for completeness and accuracy and prepared for inclusion in the SEIS.

IV. EVALUATION FINDINGS

The reviewer of information covered by this ESRP should prepare the SEIS section that lists references cited in the SEIS sections covering changes in the environmental impacts of postulated accidents during the license renewal term. The completed reference list constitutes the findings for this ESRP.

V. IMPLEMENTATION

The method described in this ESRP would be used by the staff in evaluating conformance with the Commission's regulations, except in those cases in which the applicant for license renewal proposes an acceptable alternative for complying with specified portions of the regulations.

VI. BIBLIOGRAPHY

10 CFR Part 51. *Code of Federal Regulations*, Title 10, *Energy,* Part 51, "Environmental Protection Regulations for Domestic Licensing and Related Regulatory Functions."

U.S. Nuclear Regulatory Commission (NRC). 2013. *Generic Environmental Impact Statement for License Renewal of Nuclear Plants.* NUREG-1437, Volumes 1, 2, and 3, Revision 1, Office of Nuclear Reactor Regulation, Washington, D.C.

NUREG-1555, Supplement 1

U.S. NUCLEAR REGULATORY COMMISSION

ENVIRONMENTAL STANDARD REVIEW PLAN

OFFICE OF NUCLEAR REACTOR REGULATION

6.0 ALTERNATIVES TO LICENSE RENEWAL

I. AREAS OF REVIEW

This environmental standard review plan (ESRP) provides guidance for the preparation of introductory paragraphs for the portion of the supplemental environmental impact statement (SEIS) that describes the environmental impacts of reasonable alternatives to license renewal during the renewal term.

The scope introduces the material from the reviews conducted under ESRPs 6.1 through 6.3. It includes descriptions of the reasonable alternatives to the proposed action (license renewal) discussed in the *Generic Environmental Impact Statement for License Renewal of Nuclear Plants* (GEIS; NUREG-1437, Volumes 1, 2, and 3, Revision 1) and identification of alternatives eliminated from detailed study.

Data and Information Needs

The reviewer for this ESRP requires the following information:

- list of reasonable alternatives considered by the applicant and state authorities

- list of environmental issues associated with continued plant operations during the renewal term and refurbishment

- list of alternatives eliminated from detailed study

6.0-1

II. ACCEPTANCE CRITERIA

The reviewer should ensure that the introductory paragraphs prepared under this ESRP are consistent with the intent of the following regulations:

- 10 CFR 51.45(b)(3). Alternatives to the proposed action. The discussion of alternatives shall be sufficiently complete to aid the Commission in developing and exploring, pursuant to Section 102(2)(E) of NEPA, "appropriate alternatives to recommended courses of action in any proposal which involves unresolved conflicts concerning alternative uses of available resources." To the extent practicable, the environmental impacts of license renewal and the replacement power alternatives should be presented in comparative form.

- 10 CFR 51.53(c)(2). …the applicant shall discuss in this report the environmental impacts of alternatives and any other matters discussed in 10 CFR 51.45.

- 10 CFR 51.53(c)(3)(iii). The report must contain a consideration of alternatives for reducing adverse impacts, as required by Section 51.45(c), for all Category 2 license renewal issues in Appendix B to subpart A of this part. No such consideration is required for Category 1 issues in Appendix B to subpart A of this part.

- 10 CFR 51.70(b). The draft environmental impact statement will be concise, clear, and analytic, and written in plain language with appropriate graphics.…The format provided in Section 1(a) of Appendix A of this subpart should be used. The NRC staff will independently evaluate and be responsible for the reliability of all information used in the draft environmental impact statement.

- 10 CFR 51.71(d), concerning the draft environmental impact statement will include a preliminary analysis that considers and weighs the environmental effects of the proposed action, the environmental impacts of alternatives to the proposed action, and alternatives available for reducing or avoiding adverse environmental effects.

- 10 CFR 51.95(c), concerning renewal of an operating license or combined license for a nuclear power plant. Under Parts 52 or 54 of this chapter, the Commission shall prepare an environmental impact statement, which is a supplement to the Commission's NUREG-1437, "Generic Environmental Impact Statement for License Renewal of Nuclear Plants."

- 10 CFR 51.103(a)(2) Identify all alternatives considered by the Commission in reaching the decision, state that these alternatives were included in the range of alternatives discussed in the environmental impact statement, and specify the alternative or alternatives which were considered to be environmentally preferable.

- 10 CFR Part 51, Appendix A to Subpart A of Part 51, concerning format for presentation of material in environmental impact statements

- 10 CFR 51, Appendix A(5), "Alternatives Including the Proposed Action." Identify the environmental impacts of the proposal and the alternatives in comparative form. 10 CFR 51,

Appendix A(7), concerning the environmental consequences of alternatives, including the proposed actions and any mitigating actions which may be taken. Alternatives eliminated from detailed study will be identified and a discussion of those alternatives will be confined to a brief statement of the reasons why the alternatives were eliminated. The level of information for each alternative considered in detail will reflect the depth of analysis required for sound decisionmaking.

Technical Rationale

Introductory paragraphs should provide the reader with a clear understanding of the alternatives considered and those alternatives considered for detailed analysis.

III. REVIEW PROCEDURES

Examine the applicant's ER and consider the range of reasonable alternatives. Alternatives considered are (1) build new generating capacity, (2) purchase the power from outside the system, (3) reduce power requirements through demand reduction, and (4) the no-action alternative. The reviewer should identify the criteria used in evaluating the reasonableness of the alternatives and explain which alternatives would not be considered for detailed analysis and why. A reasonable alternative must be commercially viable on a utility scale and operational prior to the expiration of the reactor's operating license or expected to become commercially viable on a utility scale and operational prior to the expiration of the reactor's operating license. The reviewer should identify the alternatives that would be carried forward for comparison with renewing the operating license of a nuclear power plant. The reviewer should discuss the extent to which these alternatives have been considered by State authorities (e.g., public service commissions and environmental, natural resource, or energy agencies).

IV. EVALUATION FINDINGS

The reviewer of information covered by this ESRP should prepare introductory paragraphs for the SEIS. The paragraph(s) should introduce the nature of the material to be presented by the reviewers of information covered by ESRP 6.1 through 6.3. The paragraph(s) should list the types of information to be presented and describe their relationships to information presented in the SEIS.

V. IMPLEMENTATION

The method described in this ESRP would be used in evaluating conformance with the Commission's regulations, except in those cases in which the applicant for license renewal proposes an acceptable alternative for complying with specified portions of the regulations.

VI. BIBLIOGRAPHY

10 CFR Part 51. *Code of Federal Regulations*, Title 10, *Energy*, Part 51, "Environmental Protection Regulations for Domestic Licensing and Related Regulatory Functions."

U.S. Nuclear Regulatory Commission. 2013. *Generic Environmental Impact Statement for License Renewal of Nuclear Plants*. NUREG-1437, Vols. 1, 2, and 3, Revision 1. Office of Nuclear Reactor Regulation, Washington, D.C.

U.S. NUCLEAR REGULATORY COMMISSION
ENVIRONMENTAL STANDARD REVIEW PLAN
OFFICE OF NUCLEAR REACTOR REGULATION

6.1 THE NO-ACTION ALTERNATIVE

I. AREAS OF REVIEW

This environmental standard review plan (ESRP) provides guidance for the analysis and assessment of potential impacts of the no-action alternative. The potential impacts of the no-action alternative have been evaluated in the *Generic Environmental Impact Statement for License Renewal of Nuclear Plants* (GEIS; NUREG-1437, Volumes 1, 2, and 3, Revision 1).

The scope of the review directed by this plan includes (1) review of the discussion of potential impacts of the no-action alternative in the GEIS, (2) identification and evaluation of new information related to potential impacts of the no-action alternative for significance, and (3) preparation of input to the supplemental environmental impact statement (SEIS).

Data and Information Needs

The reviewer for this ESRP may require the following information:

• list of environmental issues associated with the termination of nuclear power plant operations

---

II. ACCEPTANCE CRITERIA

Acceptance criteria for the review of environmental impacts of the no-action alternative are based on the relevant requirements of the following regulations:

- 10 CFR 51.45(b)(3), "Alternatives to the Proposed Action." The discussion of alternatives shall be sufficiently complete to aid the Commission in developing and exploring, pursuant to Section 102(2)(E) of NEPA, "appropriate alternatives to recommended courses of action in any proposal which involves unresolved conflicts concerning alternative uses of available resources." To the extent practicable, the environmental impacts of license renewal and the replacement power alternatives should be presented in comparative form.

- 10 CFR 51.53(c)(2). ...the applicant shall discuss in this report the environmental impacts of alternatives and any other matters discussed in 10 CFR 51.45.

- 10 CFR 51.70(b). The draft environmental impact statement will be concise, clear, and analytic, and written in plain language with appropriate graphics....The format provided in Section 1(a) of Appendix A of this subpart should be used. The NRC staff will independently evaluate and be responsible for the reliability of all information used in the draft environmental impact statement.

- 10 CFR 51.71(d). The draft environmental impact statement will include a preliminary analysis that considers and weighs the environmental effects of the proposed action, the environmental impacts of alternatives to the proposed action, and alternatives available for reducing or avoiding adverse environmental effects.

- 10 CFR Part 51, Appendix A to Subpart A of Part 51, concerning format for presentation of material in environmental impact statements

- 10 CFR 51, Appendix A to Subpart A of Part 51–4, "Purpose of and Need for Action." The alternative of no action will be discussed.

Technical Rationale

The technical rationale for evaluating the description of the no action alternative is discussed in the following paragraphs:

This ESRP examines the potential environmental impacts associated with not renewing the operating license of a nuclear power plant (i.e., the no-action alternative). 10 CFR 51, Appendix A to Subpart A of Part 51–4, "Purpose of and Need for Action," explicitly requires an analysis of the no-action alternative. If the operating license is not renewed, then the plant would cease operations at or before the end of the current operating license, and other electric-generating sources would be pursued if power were still needed.

The no-action alternative does not involve the determination of whether any power is needed or should be generated. The decision to generate power and the determination of how much power is needed are at the discretion of State and utility officials.

Not renewing the license may lead to the selection of other electric-generating sources to meet energy demands as determined by appropriate state and utility officials, conservation measures, decisions to import power, or a combination of these different outcomes. Additionally, not renewing the operating license could lead to the eventual decommissioning of the nuclear power plant and its associated impacts; these impacts are addressed separately in the *Final Generic Impact Statement on Decommissioning of Nuclear Facilities*, Volumes 1 and 2, NUREG-0586.

III. REVIEW PROCEDURES

To analyze the environmental impacts of the no-action alternative, the reviewer should complete the following steps:

1. Review the discussion of potential environmental impacts of the no-action alternative and the other alternatives in Chapter 4 of the GEIS to identify the information considered and the conclusions reached. This step establishes the basis for the evaluation of information identified by the applicant, the public, and the staff.

2. Determine if there is new information that should be evaluated. The following sources of information should be considered:

   - The applicant's ER. An applicant is required by 10 CFR 51.53(c)(3)(iv) to disclose new and significant information of environmental impacts of license renewal of which it is aware. In reviewing the applicant's ER, consider the applicant's process for discovering new information and evaluating the significance of any new information discovered.

   - Records of public scoping meetings and correspondence related to the application. Compare information presented by the public with information considered in the GEIS.

3. Determine, from the scope of environmental impacts of the no-action alternative, those that are minor and those that are likely to be sufficiently important to require detailed analysis.

   If, based on this analysis, the reviewer determines that there would be more than minor impacts, proceed to Step 4. Otherwise, if the reviewer determines that there would be no environmental impacts or that the impacts would be minor, develop a statement to this effect.

4. Analyze the environmental impacts associated with the no-action alternative, as follows:

   - Identify and evaluate the likely environmental impacts of terminating nuclear plant operations using plant-specific data.

   - Include in the analysis of impacts on environmental resource areas in the SEIS. Direct, indirect, and cumulative impacts should be considered.

Data provided in the applicant's ER include:

- the degree to which the local environmental resources would be affected by the termination of plant operations

- the significance or potential significance of environmental impacts. SMALL impacts result when no discernible change in environmental resources occurs as a result of reactor shutdown. MODERATE impacts result when there is a discernible change. LARGE impacts occur when there is substantial impact to environmental resources

- any mitigative measures for which credit is being taken to reduce environmental concerns. Supplemental data may be useful in determining the completeness of the applicant's assessment of impacts

5. Consider and evaluate potential mitigation measures or alternatives that might reduce or eliminate the adverse impacts or the disproportionate distribution of the impacts in those cases where the impacts are MODERATE or LARGE. These may have been considered in the applicant's ER.

6. Based on the results of the assessments listed above, prepare the following for the SEIS:

- a summary statement (qualitative or quantitative, as appropriate) about the degree to which environmental resources are expected to receive impacts from the no-action alternative, together with the significance of these impacts

- a discussion of the rationale or basis for conclusions supporting the degree of impact

- a discussion of any mitigative measures that would or could reduce adverse environmental impacts

IV. EVALUATION FINDINGS

The depth and extent of the information in the SEIS would be governed by the extent and significance of the effects of the no-action alternative. The reviewer should verify that sufficient information is available to meet the relevant requirements.

V. IMPLEMENTATION

The method described in this ESRP would be used in evaluating conformance with the Commission's regulations, except in those cases in which the applicant for license renewal proposes an acceptable alternative for complying with specified portions of the regulations.

VI. REFERENCES

10 CFR Part 51. *Code of Federal Regulations*, Title 10, *Energy*, Part 51, "Environmental Protection Regulations for Domestic Licensing and Related Regulatory Functions."

U.S. Nuclear Regulatory Commission (NRC). 2002. *Final Generic Impact Statement on Decommissioning of Nuclear Facilities*, Volumes 1 and 2. NUREG-0586. Washington, D.C.

U.S. Nuclear Regulatory Commission (NRC). 2013. *Generic Environmental Impact Statement for License Renewal of Nuclear Plants*. NUREG-1437, Vols. 1, 2, and 3, Revision 1. Office of Nuclear Reactor Regulation, Washington, D.C.

U.S. NUCLEAR REGULATORY COMMISSION
ENVIRONMENTAL STANDARD REVIEW PLAN
OFFICE OF NUCLEAR REACTOR REGULATION

6.2 REPLACEMENT POWER ALTERNATIVES

I. AREAS OF REVIEW

This environmental standard review plan (ESRP) provides guidance for the analysis and assessment of potential impacts of replacement power alternatives. The potential impacts of replacement power alternatives have been evaluated in the *Generic Environmental Impact Statement for License Renewal of Nuclear Plants* (GEIS; NUREG-1437, Volumes 1, 2, and 3, Revision 1).

The scope of this ESRP includes (1) review of replacement power alternative discussions in the GEIS, (2) identification and evaluation of new information about replacement power alternatives for significance, and (3) preparation of input to the supplemental environmental impact statement (SEIS).

Data and Information Needs

The reviewer for this ESRP may require the following information:

- list of reasonable replacement power alternatives considered by the applicant and State authorities

- list of environmental issues associated with continued plant operations during the renewal term and refurbishment

- list of replacement power alternatives eliminated from detailed study

---

USNRC ENVIRONMENTAL STANDARD REVIEW PLAN

Environmental standard review plans are prepared for the guidance of the Office of Nuclear Reactor Regulation staff responsible for environmental reviews for nuclear power plants. These documents are made available to the public as part of the Commission's policy to inform the nuclear industry and the general public of regulatory procedures and policies. Environmental standard review plans are not substitutes for regulatory guides or the Commission's regulations and compliance with them is not required. These supplemental environmental standard review plans are keyed to Regulatory Guide 4.2, Supplement 1, "Preparation of Environmental Reports for Nuclear Power Plant License Renewal Applications."

Published environmental standard review plans will be revised periodically, as appropriate, to accommodate comments and to reflect new information and experience.

Comments and suggestions for improvement will be considered and should be sent to the U.S. Nuclear Regulatory Commission, Office of Nuclear Reactor Regulation, Washington, DC 20555-0001.

II. ACCEPTANCE CRITERIA

Acceptance criteria for the review of environmental impacts are based on the relevant requirements of the following regulations:

- 10 CFR 51.45(b)(3), "Alternatives to the Proposed Action." The discussion of alternatives shall be sufficiently complete to aid the Commission in developing and exploring, pursuant to Section 102(2)(E) of NEPA, "appropriate alternatives to recommended courses of action in any proposal which involves unresolved conflicts concerning alternative uses of available resources." To the extent practicable, the environmental impacts of license renewal and the replacement power alternatives should be presented in comparative form.

- 10 CFR 51.53(c)(2) …the applicant shall discuss in this report the environmental impacts of alternatives and any other matters discussed in 10 CFR 51.45.

- 10 CFR 51.53(c)(3)(iii) The report must contain a consideration of alternatives for reducing adverse impacts, as required by Section 51.45(c), for all Category 2 license renewal issues in Appendix B to subpart A of this part. No such consideration is required for Category 1 issues in Appendix B to subpart A of this part.

- 10 CFR 51.70(b). The draft environmental impact statement will be concise, clear, and analytic, and written in plain language with appropriate graphics.…The format provided in Section 1(a) of Appendix A of this subpart should be used. The NRC staff will independently evaluate and be responsible for the reliability of all information used in the draft environmental impact statement.

- 10 CFR 51.71(d). The draft environmental impact statement will include a preliminary analysis that considers and weighs the environmental effects of the proposed action, the environmental impacts of alternatives to the proposed action, and alternatives available for reducing or avoiding adverse environmental effects.

- 10 CFR 51.95(c), concerning the renewal of an operating license or combined license for a nuclear power plant. Under Parts 52 or 54 of this chapter, the Commission shall prepare an environmental impact statement, which is a supplement to the Commission's NUREG-1437, "Generic Environmental Impact Statement for License Renewal of Nuclear Plants."

- 10 CFR 51.103(a)(2). Identify all alternatives considered by the Commission in reaching the decision, state that these alternatives were included in the range of alternatives discussed in the environmental impact statement, and specify the alternative or alternatives which were considered to be environmentally preferable.

- 10 CFR Part 51, Appendix A to Subpart A of Part 51, concerning format for presentation of material in environmental impact statements.

- 10 CFR 51, Appendix A(5), "Alternatives Including the Proposed Action." Identify the environmental impacts of the proposal and the alternatives in comparative form.

- 10 CFR 51, Appendix A(7), concerning the environmental consequences of alternatives, including the proposed actions and any mitigating actions which may be taken. Alternatives eliminated from detailed study will be identified and a discussion of those alternatives will be confined to a brief statement of the reasons why the alternatives were eliminated. The level of information for each alternative considered in detail will reflect the depth of analysis required for sound decisionmaking.

Technical Rationale

The technical rationale for evaluating the applicant's description of the potential replacement power alternatives to license renewal is discussed in the following paragraphs:

The GEIS does not contain any conclusions regarding the environmental impact or acceptability of alternatives to license renewal. Accordingly, the NRC must conduct an analysis of reasonable alternatives to license renewal in plant-specific environmental reviews. A reasonable alternative must be commercially viable on a utility scale and operational prior to the expiration of the reactor's operating license or expected to become commercially viable on a utility scale and operational prior to the expiration of the reactor's operating license. This ESRP examines the potential environmental impacts associated with the replacement power alternatives.

Analysis of replacement power alternatives does not involve the determination of whether any power is needed or should be generated. The decision to generate power and the determination of how much power is needed are at the discretion of State and utility officials.

III. REVIEW PROCEDURES

To analyze the environmental impact of replacement power alternatives, the reviewer should complete the following steps:

1. Review the discussion of potential environmental impacts of replacement power alternatives in Chapter 4 of the GEIS to identify the information considered and the conclusions reached. This step establishes the basis for the evaluation of information identified by the applicant, the public, and the staff.

2. Obtain information for evaluation. The following sources of information should be considered:

   - The applicant's ER. An applicant is required by 10 CFR 51.53(c)(3)(iv) to disclose new and significant information of environmental impacts of license renewal of which it is aware. In reviewing the applicant's ER, consider the applicant's process for discovering new information and evaluating the significance of any new information discovered.

   - Records of public scoping meetings and correspondence related to the application. Compare information presented by the public with information considered in the GEIS.

3. Determine, from the scope of environmental impacts of replacement power alternatives, those that are minor and those that are likely to be sufficiently important to require detailed analysis. If, based on this analysis, the reviewer determines that there would be more than minor impacts, proceed to

Step 4. Otherwise, if the reviewer determines that there would be no environmental impacts or that the impacts would be minor, develop a statement to this effect.

4. Analyze the environmental impacts associated with replacement power alternatives, as follows:

- Identify and calculate the likely environmental impacts of required replacement power alternatives including conservation and purchased or imported power, based on Chapter 4 of the GEIS, the applicant's ER, and the integrated resource plans for the area(s) or region(s) currently or (if different) likely to be served by the plant. Assume appropriate mitigation measures (e.g., emission control technologies and best management practices) for each replacement power alternative.

- Describe the environmental impacts in sufficient detail so that other environmental reviewers may evaluate the impacts of the replacement power alternatives with those of renewing the operating license. The analysis should consider the impacts on environmental resource areas in the SEIS. The analyses should include direct, indirect, and cumulative impacts. For each replacement power alternative, the analysis should identify and, to the extent possible, quantify, unavoidable adverse impacts, irreversible and irretrievable resource commitments, and tradeoffs between short-term use and long-term productivity of the environment. To the extent possible, each alternative should be analyzed on a nuclear power plant site- or region-specific basis.

Data provided in the applicant's ER include:

- the degree to which the local environmental resources would be affected by the construction and operation of replacement power alternatives

- the significance or potential significance of environmental impacts. SMALL impacts result when no discernible change in environmental resources occurs as a result of constructing and operating replacement power alternatives. MODERATE impacts result when there is a discernible change. LARGE impacts occur when there is substantial impact to environmental resources.

- any mitigative measures for which credit is being taken to reduce environmental concerns

Supplemental data obtained from other individuals and organizations may be useful in determining the completeness of the applicant's assessment of impacts.

5. Consider and evaluate potential mitigation measures or alternatives that might reduce or eliminate the adverse impacts or the disproportionate distribution of the impacts in those cases where the impacts are MODERATE or LARGE. These may have been considered in the applicant's ER.

6. Based on the results of the assessments listed above, prepare the following for the SEIS:

- a summary statement (qualitative or quantitative, as appropriate) about the degree to which environmental resources are expected to receive impacts from replacement power alternatives, together with the significance of these impacts

- a discussion of the rationale or basis for conclusions supporting the degree of impact

- a discussion of any mitigative measures that would or could reduce environmental impacts

IV. EVALUATION FINDINGS

The depth and extent of the information in the SEIS would be governed by the extent and significance of the effects of replacement power alternatives. The reviewer should verify that sufficient information is available to meet the relevant requirements.

V. IMPLEMENTATION

The method described in this ESRP would be used by the staff in evaluating conformance with the Commission's regulations, except in those cases in which the applicant for license renewal proposes an acceptable alternative for complying with specified portions of the regulations.

VI. BIBLIOGRAPHY

10 CFR Part 51. *Code of Federal Regulations*, Title 10, *Energy*, Part 51, "Environmental Protection Regulations for Domestic Licensing and Related Regulatory Functions."

U.S. Nuclear Regulatory Commission (NRC). 2013. *Generic Environmental Impact Statement* (GEIS) *for License Renewal of Nuclear Plants*. NUREG-1437, Vols. 1, 2, and 3, Revision 1. Office of Nuclear Reactor Regulation, Washington, D.C.

U.S. NUCLEAR REGULATORY COMMISSION

ENVIRONMENTAL STANDARD REVIEW PLAN

OFFICE OF NUCLEAR REACTOR REGULATION

6.3 REFERENCES

I. AREAS OF REVIEW

This environmental standard review plan (ESRP) provides guidance for the listing references cited in this chapter of the supplemental environmental impact statement (SEIS).

II. ACCEPTANCE CRITERIA

Acceptance criteria for the preparation of the reference list are based on the relevant requirements of the following regulation:

- 10 CFR 51.70(b), concerning preparation of a draft environmental impact statement (EIS) that is concise, clear, analytic, and written in plain language

III. REVIEW PROCEDURES

The environmental project manager (EPM) should contact reviewers for ESRPs 6.0 through 6.2 and compile a list of references cited in the SEIS sections that the reviewers have prepared. The citations should be checked for completeness and accuracy and prepared for inclusion in the SEIS.

IV. EVALUATION FINDINGS

The reviewer of information covered by this ESRP should prepare the SEIS section that lists references cited in the SEIS sections covering environmental impacts of continued plant operations and refurbishment. The completed reference list constitutes the findings for this ESRP.

V. IMPLEMENTATION

The method described in this ESRP would be used in evaluating conformance with the Commission's regulations, except in those cases in which the applicant for license renewal proposes an acceptable alternative for complying with specified portions of the regulations.

VI. BIBLIOGRAPHY

10 CFR Part 51. *Code of Federal Regulations*, Title 10, *Energy,* Part 51, "Environmental Protection Regulations for Domestic Licensing and Related Regulatory Functions."

U.S. Nuclear Regulatory Commission (NRC). 2013. *Generic Environmental Impact Statement for License Renewal of Nuclear Plants.* NUREG-1437, Vols. 1, 2, and 3, Revision 1. Washington, D.C.

U.S. NUCLEAR REGULATORY COMMISSION
ENVIRONMENTAL STANDARD
REVIEW PLAN
OFFICE OF NUCLEAR REACTOR REGULATION

7.0 SUMMARY AND CONCLUSIONS

I. AREAS OF REVIEW

This environmental standard review plan (ESRP) provides guidance on preparing this chapter of the supplemental environmental impact statement (SEIS) that integrates the conclusions for issues designated Category 1 or resolved Category 2 in the *Generic Environmental Impact Statement for License Renewal of Nuclear Plants* (GEIS; NUREG-1437, Volumes 1, 2, and 3, Revision 1), information developed for those open Category 2 issues applicable to the plant, and significant new information. The chapter must conclude whether the adverse environmental impacts of license renewal are so great that preserving the option of license renewal for energy planning decisionmakers would be unreasonable.

The scope includes (1) review of the impact analyses prepared for the SEIS, (2) evaluation of the cumulative impacts associated with continued plant operations during the license renewal term and refurbishment, (3) review of the discussions of the environmental impacts of alternatives, (4) comparison of the environmental impacts of license renewal with the environmental impacts of the alternatives, and (5) preparation of input to the SEIS.

The SEIS input should (1) identify adverse environmental impacts that are unavoidable, (2) identify commitments of resources that are irreversible and irretrievable, and (3) discuss the effects of short-term use on maintenance and long-term productivity of the environment.

<u>Data and Information Needs</u>

The types of data and information needed would be affected by nuclear power plant site- and plant-specific factors. The following data or information may be needed:

- the discussion of environmental impacts of license renewal in the GEIS

- the discussion of plant-specific environmental impacts of license renewal in the applicant's environmental report (ER)

- the summary of environmental impact analyses conducted for the SEIS

II. ACCEPTANCE CRITERIA

Acceptance criteria for the preparation of the summary and conclusions are based on the relevant requirements of the following regulations:

- 10 CFR 51.70(a), concerning preparation of a draft environmental impact statement (EIS)

- 10 CFR 51.70(b), concerning a concise, clear, analytic EIS written in plain language

- 10 CFR 51.71(d), concerning including a preliminary analysis that considers and weighs the environmental effects of the proposed action, the environmental impacts of alternatives to the proposed action, and alternatives available for reducing or avoiding adverse environmental effects

- 10 CFR 51.71(d), concerning including a preliminary recommendation by the NRC staff respecting the proposed action reached after considering the environmental effects of the proposed action and reasonable alternatives

- 10 CFR 51.95(c)(4), concerning preparation of a final EIS

- 10 CFR 51.95(c)(4), concerning including the NRC staff recommendation regarding the environmental acceptability of the license renewal action that integrates the conclusions, as amplified by the supporting information in the generic EIS, for issues designated Category 1 or resolved Category 2, information developed for those open Category 2 issues applicable to the plant, and any significant new information.

<u>Technical Rationale</u>

The SEIS must include NRC staff recommendations regarding the environmental acceptability of the proposed action. In making these recommendations, the staff is required to integrate the conclusions from the GEIS, plant-specific impact analyses, and any significant new information. This ESRP summarizes the environmental impacts of the proposed action, comparison of the environmental impacts of the proposed action with the impact of the alternatives, and the staff recommendations.

III. REVIEW PROCEDURES

The environmental project manager (EPM) is responsible for the preparation of the SEIS summary and conclusions section. The summary and conclusions section should be sufficiently complete that a person reading this section would understand:

- the purpose of and need for the proposed action

- the NEPA process and NRC's environmental review leading to the preparation of the SEIS

- the environmental impacts of renewing the operating license

- the environmental impacts of alternatives to renewing the operating license

- staff conclusions and recommendations

Suggested steps for the preparation of the summary and conclusions chapter of the SEIS are as follows:

1. Prepare introductory paragraphs for the Summary and Conclusions chapter.

2. Prepare a table that summarizes the findings of the environmental impacts presented in the SEIS. The summary and conclusions table should list of the environmental impacts of license renewal and alternatives to license renewal (including no-action) and state the level of significance of each impact. This table should be organized by area of environmental concern.

3. The EPM should also consider the list of unavoidable adverse impacts and the list of irreversible and irretrievable resource commitments, and draw conclusions related to effects of short-term commitments on maintenance and long-term productivity of the environment. The final lists of unavoidable adverse impacts and irreversible and irretrievable resource commitments and a discussion of the effects of short-term use on maintenance and long-term productivity of the environment should also be included in the SEIS.

4. Prepare input to the SEIS summary and conclusions chapter.

IV. EVALUATION FINDINGS

The EPM prepares the SEIS section that presents (1) the overall summary of the environmental impacts of license renewal and alternatives to license renewal (including no-action), and (2) the staff recommendations regarding license renewal. The overall summary should be presented in tabular form. The contents of the table are described in the Review Procedures section. The staff recommendation should be stated in terms consistent with the wording of 10 CFR 51.95(c)(4).

V. IMPLEMENTATION

The method described in this ESRP would be used by the staff in evaluating conformance with the Commission's regulations, except in those cases in which the applicant for license renewal proposes an acceptable alternative for complying with specified portions of the regulations.

VI. BIBLIOGRAPHY

10 CFR Part 51. *Code of Federal Regulations*, Title 10, *Energy,* Part 51, "Environmental Protection Regulations for Domestic Licensing and Related Regulatory Functions."

U.S. Nuclear Regulatory Commission (NRC). 2013. *Generic Environmental Impact Statement for License Renewal of Nuclear Plants.* NUREG-1437, Vols. 1, 2, and 3, Revision 1. Office of Nuclear Reactor Regulation, Washington, D.C.

NRC FORM 335
(12-2010)
NRCMD 3.7

U.S. NUCLEAR REGULATORY COMMISSION

BIBLIOGRAPHIC DATA SHEET

(See instructions on the reverse)

1. REPORT NUMBER
(Assigned by NRC, Add Vol., Supp., Rev., and Addendum Numbers, if any.)

NUREG-1555, Supplement 1,
Revision 1
FINAL

2. TITLE AND SUBTITLE

Standard Review Plans for Environmental Reviews for Nuclear Power Plants
Supplement 1: Operating License Renewal

3. DATE REPORT PUBLISHED

MONTH	YEAR
June	2013

4. FIN OR GRANT NUMBER

5. AUTHOR(S)

6. TYPE OF REPORT

Technical

7. PERIOD COVERED (Inclusive Dates)

8. PERFORMING ORGANIZATION - NAME AND ADDRESS (If NRC, provide Division, Office or Region, U. S. Nuclear Regulatory Commission, and mailing address; if contractor, provide name and mailing address.)

Division of License Renewal
Office of Nuclear Reactor Regulation
U.S. Nuclear Regulatory Commission
Washington, D.C. 20555-0001

9. SPONSORING ORGANIZATION - NAME AND ADDRESS (If NRC, type "Same as above", if contractor, provide NRC Division, Office or Region, U. S. Nuclear Regulatory Commission, and mailing address.)

Same as 8 above

10. SUPPLEMENTARY NOTES

11. ABSTRACT (200 words or less)

This document provides guidance to U.S. Nuclear Regulatory Commission staff in implementing the provisions in Title 10 of the Code of Federal Regulations Part 51 (10 CFR Part 51), "Environmental Protection Regulations for Domestic Licensing and Related Regulatory Functions" when conducting an environmental review for the renewal of a nuclear power plant operating license(s). This standard review plan guides the staff in preparing a plant-specific supplemental environmental impact statement to NUREG-1437, Revision 1, Generic Environmental Impact Statement for License Renewal of Nuclear Plants. This document supplements NUREG-1555, Standard Review Plans for Environmental Reviews for Nuclear Power Plants, which provides guidance for the environmental reviews of construction permits, initial operating licenses, early site permits, and combined licenses for new nuclear power plants.

12. KEY WORDS/DESCRIPTORS (List words or phrases that will assist researchers in locating the report.)

Standard Review Plans for Environmental Reviews for Nuclear Power Plants
ESRP
NUREG-1555, Supplement 1, Revision 1
National Environmental Policy Act
NEPA
License Renewal

13. AVAILABILITY STATEMENT

unlimited

14. SECURITY CLASSIFICATION

(This Page)

unclassified

(This Report)

unclassified

15. NUMBER OF PAGES

16. PRICE

NRC FORM 335 (12-2010)

Printed
on recycled
paper

Federal Recycling Program

UNITED STATES
NUCLEAR REGULATORY COMMISSION
WASHINGTON, DC 20555-0001

OFFICIAL BUSINESS

NUREG-1555,
Supplement 1, Rev. 1

Standard Review Plans for Environmental Reviews
for Nuclear Power Plants

June 2013